DEEPSEA MINING

Selected Papers from a Series of Seminars Held at the
Massachusetts Institute of Technology in December
1978 and January 1979

Edited by Judith T. Kildow

The MIT Press
Cambridge, Massachusetts, and London, England

This volume was prepared with the support of a
National Science Foundation Grant, number OCE-7815700.
The opinions, findings, conclusions, and recommenda-
tions expressed in this publication are those of the
authors and do not necessarily reflect the views of
the National Science Foundation.

Library of Congress Cataloging in Publication Data
Main entry under title:

Deepsea mining.

    Includes bibliographies.
    1.   Ocean mining--Congresses.   I.   Kildow, Judith
Tegger.
TN291.5.D43   333.8'09162   79-23633
ISBN 0-262-11-075-X

CONTENTS

Contents

The papers in this volume were among those presented at a series of four seminars held at the Massachusetts Institute of Technology during December 1978 and January 1979. The seminars--jointly sponsored by the departments of ocean engineering, materials science and engineering, and the Center for Policy Alternatives, and supported by a grant from the National Science Foundation--were designed to examine several policy issues related specifically to deepsea mining but also more broadly to international resource management. It is commonly agreed that the availability of a valuable mineral resource in a seabed territory delimited as belonging to the heritage of mankind exacerbates a set of older, unresolved national and international policy problems by adding a group of newer, even more poorly understood problems. The newer problems reflect fundamental changes in the international political and resource supply systems; and the concurrence of these interlinked dynamic forces has created an awesome challenge for policy makers throughout the world.

Before discussing the individual papers, I should remark on the absence of concentrated discussion of one crucial topic: environmental impacts. Both the harvesting and the processing of the resource present potential problems. Unfortunately, despite the attempt by the Department of Commerce to do a small-scale on-site assessment for harvesting in its DOMES project, the data base for the study of these problems remains shaky at best. Given this lack, which will be remedied only when we have a better idea of the shape of a full-scale production process, we felt it best to exclude the topic rather than discuss it incompletely and prematurely. This should not, of course, be taken as an indication of lack of interest in these problems; inevitably, environmental considerations will have a major effect on the shape of policy.

The first paper, by Judith T. Kildow and Vinod K. Dar, provides an overview of the deepsea mining con-

troversy in both national and international frameworks.
The paper stresses the unusual nature of the resource
and the problems it presents to industries and governments; but it also shows that many of the problems
are general and must be considered in the broader context of global resource utilization and management
and the inevitable changes that are coming in the
international system.

The second section opens with a paper by Jane Z.
Frazer of the Scripps Institution of Oceanography,
who maintains the central data bank in the public
sector on the abundance and grade of the deepsea
resource. She describes the current data base and
some of the controversies that have come up in its
collection and interpretation.

The paper by Lance Antrim contrasts the availability of the metals found in the manganese nodules
with land-based supplies. He thereby demonstrates
the relative importance of deepsea mining over time
in supplementing the worlds supply of the four
principal metals found in these deposits: nickel,
cobalt, copper, and manganese.

The next two papers provide estimates of the net
effects of deepsea mining on the markets for the
four metals and on the costs to the United States,
taking into consideration forecasts of supply and
demand and possible technological changes. The
paper by Bernard J. Reddy and Joel P. Clark assesses
the potential market impacts, trying to determine
where the benefits and disbenefits of deepsea mining
will fall. James C. Burrows then examines some of
the positive results of seabed mining for the United
States, including reduced probability and severity
of cartelization, reduced rate of depletion of land-
based reserves, and increased military and political
security. Through this analysis, Burrows derives an
approximate net value of the resource to the United
States over time.

The final paper in the second section discusses
a potentially significant technological breakthrough

that could affect the global market for manganese
and thus change the configuration of policy issues
considerably. Nicholas J. Grant, who has been working on this project for some time, indicates the
extent to which technological changes can influence
the entire market system for many of these metals
and as a consequence affect the politics of the
resources as well.

The third section opens with two differing
assessments of U.S. policy on deepsea mining. Arthur
Kobler, a staff member of the Department of State
Office of Commodities who has worked with the Common
Fund in the UNCTAD discussions, outlines the government's rationale for its positions and discusses some
of the problems therein. Richard G. Darman, a former
member of the U.S. delegation to the Law of the Sea
Conference, presents a more critical analysis, giving
the reader a broader time perspective and illumina-
ting some interesting and controversial patterns that
have emerged.

A Third World perspective is offered next by A.
O. Adede of Kenya, a staff member of the U.N. Legal
Office. Adede describes a broad range of issues that
he feels to be important in negotiating the Seabed
Authority that would regulate the development of the
resource; he emphasizes the points of controversy
that remain.

The next three papers provide an array of industrial perspective. In the first, Burton H. Klein,
an economist from the California Institute of
Technology, describes some of the structural problems
of the large industries that must provide the
innovative pioneering for the deepsea mining effort.
The paper by J. A. Agarwal, who has played a central
role in the development of the processing technologies proposed by the Kennecott Copper Company for its
deepsea mining effort, then offers a pragmatic
technical viewpoint. The final paper in the section
was written by John E. Flipse, now a professor at
Texas A&M and formerly president of Deepsea Ventures,

one of the first consortia formed for the purpose of developing the manganese nodules. Flipse takes a sharp look into the future of the industry and identifies some of the obstacles ahead.

The final paper in the book summarizes the key points raised in the seminars, outlining both areas of consensus and outstanding issues where conflicting perspectives and values hinder consensus.

The seminars were conducted principally by myself and my colleague Joel Clark. I would also like to express gratitude for the valuable assistance provided by our steering committee, which included Richard Baxter, Michael Bever, James Burrows, Gordon Christenson, Marne Dubs, Ira Dyer, Herbert Holloman, Alan Kaufman, Amor Lane, Walter Owen, Robert Seamans, and Maxwell Morton, and by those who participated in the seminars and contributed to the discussions that followed the papers. I am also grateful for the administrative and editorial assistance of Heinz Stubblefield and Holly Altman.

A final acknowledgment of appreciation goes to the Office of Marine Affairs of the IDOE in the National Science Foundation for its financial support, which made this volume and the seminar discussions possible.

Judith T. Kildow
Cambridge, Massachusetts

A. O. Adede
Office of Legal Affairs
U.N. Secretariat
New York, NY

J. A. Agarwal
Kennecott Copper
Corporation
Lexington, MA

Lance Antrim
Office of Ocean,
Resource, and
Scientific Policy
Coordination
U.S. Department of
Commerce
Washington, D.C.

James C. Burrows
Charles River Associates
Boston, MA

Joel P. Clark
Department of Materials
Science and Engineering
MIT
Cambridge, MA

Vinod K. Dar
Resource Planning
Associates
Washington, D.C.

Richard G. Darman
Kennedy School of
Government
Harvard University
Cambridge, MA

John E. Flipse
Department of Ocean
Engineering
Texas A&M University
College Station, TX

Jane Z. Frazer
Scripps Institution of
Oceanography
University of California
at San Diego
La Jolla, CA

Nicholas J. Grant
Department of Materials
Science and Engineering
MIT
Cambridge, MA

Judith T. Kildow
Department of Ocean
Engineering
MIT
Cambridge, MA

Burton H. Klein
Department of Humanities
and Social Sciences
California Institute
of Technology
Pasadena, CA

Arthur Kobler
Commodities Office
Department of State
Washington, D.C.

Bernard J. Reddy
Charles River Associates
Boston, MA

PART I

CHANGING INSTITUTIONS AND RESOURCE CONDITIONS

INTRODUCTION TO AN UNUSUAL RESOURCE MANAGEMENT PROBLEM

Judith T. Kildow and Vinod K. Dar

INTRODUCTION

Concern over the availability of stable resource
supplies to fuel the U.S. economy is by no means a
new phenomenon. What is new, however, are the
political concerns that have been added to the
traditional problem of the gradual depletion of
supplies. The United States has become increasingly
dependent on imports and is now quite vulnerable to
foreign political and economic actions. Within the
past few years, for example, we have seen:

1. an exertion of market power by foreign bauxite
producers;

2. political manipulation of the oil market in the
1973 embargo;

3. a violent conflict disrupting cobalt and copper
mining in Kolwezi, Zaire, causing a rapid escalation
of the free market price of cobalt from $12.50/kg to
over $60/kg.

4. labor strife in Canada causing serious shortages
of nickel and a sharp rise in price;

5. manipulation of the international chromium market
by the Soviet Union, leading to a tripling in the
price of its ore and sharp increases in the prices
charged by other producers (Turkey, Sourth Africa);

6. market pressure by foreign copper producers who
increased their share of the U.S. market from 5% in
1976 to 20% in 1978 while keeping the world price
below the level necessary for a financially healthy
domestic industry.

A number of changes in world conditions have increased the likelihood that such disruptive actions will occur even more frequently in the future. These changes can be classified as strategic, structural and attitudinal. Strategic changes include the emergence of resource powers such as Australia, Brazil, Canada, South Africa (for chromium in particular), and Zaire and Zambia (for cobalt); the Soviet push to dominate major African and Middle Eastern Resource-producing areas; and attempts by some developing countries to expand their market shares at the expense of U.S., Canadian, and Western European mining firms by pursuing policies of revenue and employment maximization rather than profit maximazation.

The structural changes result mainly from the process of decolonization, which has removed a number of major resource reserves from the control of Western economies. This has often been accomplished through the nationalization of Western mining investments. Another structural change has been the consolidation of efforts on the part of resource producers pursuing a common interest--the extraction of economic rent from consumers--which has resulted in new alignments and new coalitions in international bargaining forums. (On the other hand, a decaying of traditional coalitions seems to be taking place among developed countries as they find themselves increasingly in competition for the same shrinking pool of resources.)

Finally, a pervasive attitudinal change toward the global distribution of economic rents among developing and developed countries--typified by the debate over the new economic order--has lent legitimacy to attempts by resource producers to cartelize world markets and to seek means of stabilizing their revenues. These attitudinal changes are multifaceted and should not, as is too often done, be reduced to caricature. Some producing nations see their newfound power as a means of attaining equality with their industrialized

customers by raising themselves to a respectable
competitive position in the world market. Others
see an opportunity to institute fundamental changes
in the functioning of the world market system based
on a new ideological outlook and a new set of
objectives. While the latter states are consciously
in conflict with the industrialized states, and so
may provide a legitimate basis for concern in those
states, the former group finds the general distrust
manifested by the developed countries discouraging
and frustrating. These more moderate resource
producers may hold the key to a middle road, although
current debates are still marked by extreme swings
of the pendulum. (See, for example, the 1977 and
1978 informal composite negotiating texts of the U.N.
Law of the Sea Conference.)

     Several recent domestic trends have complicated
the problem of U.S. resource security. First, new
environmental and safety laws are finally forcing
U.S. firms to internalize the social costs of the
health and safety risks they pose to their workers
and also of the degradation of land, water, and air
quality that they cause. One might, of course, argue
that this forced convergence between the social and
private marginal costs of production is in the
national interest; but it does practically reduce
the economic rents available to mining corporations
and hence inhibits economic growth, adds to unemploy-
ment problems, and may contribute to inflation.

     Second is the problem of the chaotic state of
the national regulatory system for resource management. As resource security has emerged as a policy
issue, more and more bureaucratic agencies have
become involved in an uncontrolled decision-making
structure. For example, the bureaucrats in the
Department of the Interior, who have heretofore had
domestic mandates only, have now become international
resource managers, unofficially--and, indeed, without
any recognized structural liaisons--augmenting the
operations of the Department of State (see Hopkins,

1976). Given this proliferation of interested
parties, the problems of policy coordination and
central decision-making are inevitably exacerbated
by the classic problems of bureaucratic rivalry.

Finally, there has been a general recognition of
the fact that the economics of most resources are
inextricably intertwined, and that resource managers
must therefore take a broad view of their tasks.
(The interdependence of resources was shown most
effectively by the major reverberations that accompanied the 1973 oil embargo.) This more enlightened
view has dated must of the existing piecemeal legislation and also much of the existing bureaucratic
decision-making structure.

RESOURCES AND POLITICAL POLICY
The control of resources is an essential element of
political power. The desire for such control has,
of course, motivated extraterritorial excursions
throughout history; but the current world situation
seems to have increased the importance of this aspect
of the multidimensional equations of power.

The combination of changing perceptions and
evolving geopolitical resource strategies is occurring at a time when a potentially major new resource--
the ferromanganese deposits on the seabed--is emerging
into importance because of the development of
appropriate technologies. This situation demands the
formulation of a U.S. government response that
transcends conventional organizational structures,
since it involves both foreign and domestic policy
and requires a blending of the roles of leader and
follower.

The ferromanganese deposits contain several
minerals in amounts that vary according to location.
This special characteristic causes two problems.
First, it requires a linkage in the markets of all
the minerals that will be mined, which may in turn
imply a convergence in the prices of those minerals.
Second, it may cause production problems to the

extent that the proportions in which the minerals are
found in the deposits differ from the proportionate
sizes of their world markets. Clearly there is a
need for simultaneous production, pricing, and marketing decisions relating to all the product minerals.

A second fundamental problem is that of legal
ownership of the resource. All current international
actions seem predicated on the view that the resource
is supranational (that is, its ownership is vested in
all nations) rather than transnational (in which case
its ownership would be vested in no single nation).
In the case of a transnational resource, any nation
could pursue unilateral actions without violating
the rights of other nations. In the case of a
supranational resource, unilateral actions auto-
matically impinge on the rights of other states.
The very existence of the Law of the Sea negotiations
implies the acceptance of some form of supranationality for the resource. Since the United States has
no defined policy for either transnational or supranational resources, the implications for precedent-
setting in this case are awesome. (The Antarctic
is another rich store of resources that seems to be
in transition from a somewhat unstable transnational
status to one of supranationality.)

Finally, the policy adopted for seabed minerals
must ·reflect the growing recognition of the fallacy
of the trichotomous division conventionally made
among land, sea, and space resources. What is needed
is a coordinated decision-making framework that links
not only different resources, but also all possible
sources of any given resource.

AN UNUSUAL U.S. NATIONAL PROBLEM
A number of conceptual and structural problems have
become evident in the resource policy-making process
of the United States. These difficulties emanate
in large part from the uncoordinated integration of
foreign and domestic policy systems that has occurred
in response to the increasingly complicated world

resource picture. Because domestic and foreign
policies often have differing broad objectives, and
because the two policy-making structures are still
artifically separated, it is possible to set up a
domestic resource policy objective and then have it
directly or indirectly undermined by a concurrent
foreign policy objective, resource or otherwise, in
the formulation of which policy-makers failed to
consider the implications of domestic policy. (Of
course, the reverse situation can occur as well.)

While foreign policy experts and resource managers
have always recognized the random nature of the
distribution of the world's resources, the changing
international political and economic environment has
recently lent more importance to this phenomenon.
The lack of congruence between the maps of producers
and consumers of natural resources gives power to
those nations that command large resource stocks.
To the extent that the existence of this power is
recognized by the U.S. government, securing access
to foreign sources of supply becomes an important
component of U.S. foreign policy, requiring the
addition of a new dimension of "resource diplomacy."

Securing access to resources is only one
component of foreign policy; the overriding objective
is the maximization of political power. It may be
expected, then, that if and when the securing of
resources comes into conflict with the objective of
maximizing political power, other foreign policies
may dominate the international component of resource
policy (consider, for example, the U.S. policy of
nonintervention in the recent Zaire conflict, despite
U.S. dependence on cobalt from Zaire).

As the spatial aspect of resources spanning
national boundaries influences and is influenced by
foreign policies, the temporal allocation of the
consumption of resources in the United States affects
and is affected by domestic policies with regard to
resource producers and consumers. Since the role of
the federal government seems to be to mediate between

An Unusual Resource Management Problem

producers and consumers to effect a balan
competing interests, its policies have a
domestic producer and consumer surpluses
example, tariff policies that limit foreign imp
maintain producer surpluses above the levels they
would have attained in the absence of trade
restraints and lower consumer surpluses. Presumably,
foreign imports are restricted because they are
cheaper, so that domestic consumers pay a higher
price than they would have had to pay otherwise.

On the other hand, policies aimed at price
control, as in the case of natural gas, yield a gain
in consumer surpluses but a loss of rent to producers
that might have contributed to economic growth
(through investments in additional productive capacity
and higher employment). Price controls may also lead
to a transfer of wealth to foreign producers since
they tend to increase levels of imports above what
they need to be.

When international markets are distorted by
national policy interventions, resource allocations
become inefficient. The ultimate example of such a
distortion would be overconsumption from a global
point of view of a resource that is abundant in a
certain part of the world but cannot be sold to
consumers in other parts of the world where the same
resource is scarce. In short, any interference with
the function of a market (other factors being equal)
causes the shadow price of a resource to be relatively
high in the area of scarcity and relatively low in
the area of abundance.

In the case of energy resources, the low price of
petroleum in the United States over an extended
period of time led to profligate use of this raw
material. When shortages appeared probable and
access grew more difficult, a crisis was fomented in
which the need to change the use and allocation of
the resource became a catalyst for changes in
domestic and foreign policy-making systems throughout
the world.

Deepsea mining provides a good case study of the structural and substantive dilemmas of current U.S. resource policy. The lack of direction or defined mission for the numerous agencies and committees involved is exacerbated by a veil of rivalries and other uncertainties.

A major problem is that there is no legally defined lead agency to oversee or in any way make policy for deepsea mining. Anticipating national legislation and perhaps an international treaty that will define a regulatory regime, several departments have established offices to oversee their interests in seabed mining. The Department of the Interior's Ocean Mining Administration, established in 1974, and the Department of Commerce's Office of Ocean Minerals, established about a year later, have each been suggested as the lead office in legislation introduced into Congress over the past few years. The Office of Ocean Minerals is the choice of the House Merchant Marine and Fisheries Committee, while the Ocean Mining Administration is favored by the Senate Interior and Insular Affairs Committee. (This particular rivalry might have been resolved if the NOAA, the agency that houses the Office of Ocean Minerals, had been transferred to a transformed Department of the Interior as envisioned under a recently withdrawn reorganization plan.)

The list of agencies that have taken an interest in the issue goes on and on. The National Science Foundation's IDOE program has had a manganese nodule project under way since 1970, though much of this research has not yet been made public. The Department of the Interior's Bureau of Mines has funded major research contracts in the area. And numerous other agencies have done or are considering doing in-house studies, among them the Department of State's Legal Office; Office of Economic Affairs, and Commodities Office; the Treasury Department's Office of International Affairs; the Defense Department's Office of International Security Affairs;

and the Congressional Office of Technology
Assessment. And finally, the Legislative Reference
Service of the Library of Congress has published
numerous studies to aid the workings of various
congressional committees, and the National Academy
of Sciences and Engineering has also sponsored work
and published reports on the topic. All this
activity without any legislative mandate, without
any deepsea mining actually taking place, and without
any coordination.

The U.S. government is, of course, not the only
sponsor of such efforts. The industry itself has its
own meetings, reports, and conferences; the U.N.
Secretariat's Office of Ocean Economics has been
sponsoring work on the subject for at least six
years; the U.N. Legal Office has special people
assigned to the topic; the European Economic Community
has sponsored conferences; and individual nations
have also sponsored research and conferences.
Perhaps the only continuity has been an informal
network of a few individuals whose interests and
attendance at all these meetings draw them together.

Why all the activity focused on this one topic?
And why do the problems seem so difficult to solve?
(Congress has been considering various pieces of
legislation for over six years, an international
treaty has been debated for the last ten years, and
corporate interest goes back in some cases more than
fifteen years.) Are there, then, unusual conditions
prevailing that lift the problems beyond the scope of
current governmental structures?

The next two sections will deal with aspects of
the problem that are unusual from the point of view
of the international community and industry. Here
we shall deal briefly with the point of view of the
U.S. government. In general, however, we might say
that deepsea mining is unusual in that it combines
the standard policy problems associated with the
major resource industries and related technologically
oriented businesses with novel conditions, both

technical and legal, that are not yet well understood
by anyone. These new conditions might have been more
easily dealt with had the more standard problems
been resolved earlier. Consider, for example, the
much-discussed gap between "have" and "have not"
nations. This problem existed long before the advent
of the technology that made deepsea mining feasible,
but it has now been absorbed into this issue as
Third World nations have come to base their hopes
for a major shift in income allocations on new
agreements relating to natural resources, technology
transfer, and management.

For the United States, the problem is that
deepsea mining falls into several categories of
policy issues that are currently under heavy
scrutiny by the federal government: (1) nonfuel
minerals policy, (2) U.S. interests in the ocean,
(3) the U.S. role in the Third World, and (4) the
most effective organization of government to handle
resource-related problems. The policy that is
eventually worked out for seabed mining will
inevitably reflect the outcomes of policy decisions
in all of these areas. And the effectiveness of the
resolution will depend in no small part on the degree
of consistency and consensus on goals achieved among
the groups involved in these policy areas.

Meanwhile, in the international arena, the
United States has so far failed to assess its
competitive strengths and to set its negotiating
priorities on the basis of using and maintaining
those strengths. With regard to the Law of the Sea
negotiations, the greatest strength of the United
States is its civilian technology, but it has bar-
gained that strength away in favor of a military
option that plays to the greatest strength of its
chief competitor, the Soviet Union.

Without an assessment of its own potential
strengths and of its ultimate role in the evolving
international system, the United States finds itself
caught in a situation where economic security

problems are increasing, traditional military
problems are not going away, and concessions on one
issue are perceived as being necessary to save the
other.

At this time, moreover, another fundamental
domestic problem seems to dominate national policy-
making. The issue--the appropriate role for the
federal government in private-sector activities--has
manifested itself strongly in both fuel and nonfuel
resource areas. The scandals and deceptions that
have tarnished the image of the petroleum companies,
for example, have engendered much criticism of the
breadth of independence permitted strategically
important industries. Distrust in the methods of
private enterprise and a belief that there is a
growing gap between public and private sector values
and goals have led to pressures on U.S. regulators
for more control. Yet the companies themselves and
many economists claim that too much control will
discourage private investment and create disincentives
for innovation and growth. The search for a middle
way will clearly be difficult. The lack of a
coherent policy framework to deal with this basic
problem on the national front, however, leaves what
is essentially a policy vacuum on the international
front in most areas affected, among which one must
certainly count deepsea mining.

AN UNUSUAL INTERNATIONAL PROBLEM
The relative bargaining powers of states in the
debate over ocean bed resources are influenced by
economic and political factors. The economic issue
is the dispute over the division of rent. The
political issue is the dispute over the sharing of
control and power. Both these considerations are
manifestations of the struggle over the distribution
of benefits to be won by developing ocean bed
resources.

As power centers fragment and multiply, no
ideological bloc can claim resource self-sufficiency.

The desire to control resources is motivated by the economic utility derived from consumption as well as by the political utility that command over a scarce commodity generates. The tendency to use resources as a political weapon is evident: OPEC uses oil and the United States uses enriched uranium and military technology to secure ideological concessions that are only remotely related to the appropriation of rent. While the marginal rate of substitution between political and economic utility cannot be well defined, it clearly exists.

The United States, the Soviet Union, the nations of the EEC and COMECON, Japan, OPEC, and the more advanced developing countries such as Brazil and China, have begun to realize that they are all competing for the same global pool of resources. This emerging resource competition has added a potent dimension to the traditional East-West and North-South conflicts, leading to shifting coalitions of resource-supplying countries versus resource-consuming countries, depending on the international forum and commodity under discussion. The Soviet Union seems most active at present in linking control over resources with political control. A major component of the Soviet Union's geopolitical strategy seems to be to spread its influence over areas that are major suppliers of minerals to the West. The continuing Soviet incursion in Africa is a threat to the West because it carries the potential for a loss of control over the vital oil routes from the Middle East and over the sources of supply for at least ten important minerals: 62.5% of the world's output of diamonds, 60.8% of gold, 50% of vanadium, 45.7% of the platinum group metals, 36.2% of chromite, 35.3% of cobalt, 34.4% of manganese, 18% of copper, 13.7% of uranium oxide, and 13.3% of asbestos are derived from southern Africa. If the Soviets succeed in acquiring some degree of control over the supply of these minerals, they would be able to manipulate world markets. Relatively small adjustments in

supply can have a dramatic impact on prices owing to
the low short-run elasticity of demand for these
minerals. The Soviets might use the threat of price
increases to apply political pressure to the West; or
they might simply allow prices to rise for the effect
this would have on the West's economy.

The West could conceivably respond by denying the
Soviets trade privileges, food, and technology, or by
developing other sources of supply for at least some
minerals. One of these sources might be ocean bed
ferromanganese deposits.

The major nations of the world, however, are no
longer the exclusive centers of power. New actors--
private, national, international, and transnational--
are emerging, resulting in diffusion of power. This
fragmentation of power prevents even the great powers
from pursuing strategies of independence through
unilateral exploitation of ocean bed resources.
Growing interdependence will determine the nature of
international transactions, and despite the myriad
conflicts interdependence will engender, the cost of
using force to resolve these conflicts will greatly
exceed the cost of cooperative bargaining.[1] The
resolution of conflicts generated by interdependence
rests on the presumption that the mutually dependent
actors can adjudicate their differences by referring
to a larger common good, but unfortunately this calls
for a supranational institution, the likes of which
do not exist, since no institution at present has the
legitimacy and power to speak out for the common
interests of mankind.

The prime determinants of the bargaining strengths
of nations appear to be international egalitarianism
and international interdependence. These subsume the
concerns over producer and consumer cartels, unilateral exploitation of ocean bed and other supranational resources by nations that possess the
requisite technology, capital, and organizational
skills, the demands of the new economic order, and
the host of resource transfer arrangements lumped

under revenue stabilization schemes. While a
detailed examination of each of these factors would
be a mere repetition of arguments presented at length
in the literature, the broader issues of egalitarianism and interdependence merit some elucidation
since they are vitally associated with the current
political and economic debate.

With reference to deep oceanbed minerals the
main political components within the larger, encompassing concerns of international interdependence and
international egalitarianism are:

1. The freedom of the major maritime powers to
conduct scientific investigations and carry out naval
reconnaissance and patrols.

2. The unilateral appropriation of the deep oceans
by the major powers.

3. The compulsory transfer of ocean mining technology from the nations of the North to those of
the South.

4. Quotas on the production of minerals from marine
ferromanganese deposits to protect the interests of
existing exporters, often from the South.

5. Access, largely by the nations of the North, to
potentially stable sources of important minerals
such as copper, cobalt, and nickel.

6. The decline of military power as a factor in the
settlement of international disputes and the con-
comitant rise of economic power (in the form of
control over oil and nonfuel minerals) as a source
of leverage.

7. The distribution of the benefits from marine
minerals development (beyond the 200-mile economic
zone) within the general framework of the new

economic order, entailing a net transfer of resources from the North to the South.

8. The development of ferromanganese nodules through private capital supplied by international mining consortia drawn from Western Europe, Japan, and North America, or through international public capital controlled by a transnational seabed regime dominated by the "Group of 77" and initially funded by the nations of the North and perhaps OPEC.

9. The integration of deep seabed minerals produc- tion with a general scheme for commodity price stabilization and international commodity agreements involving producers and consumers.

The international regime of nations has been blessed with neither great order nor equality. The appropriation of rent, both economic and political, from the use of the planet's resources has always had a certain element of anarchy and oligarchy since those conditions have been the prime characteristics of the global system (see Aron, 1968, p. 160). Recent years, however, have seen growing demands for greater equality among nations, an equality trans- cending the traditional inequalities based on size, military power, resource endowments, population, and technological stock. This new egalitarianism has found its principal institutional expression in the United Nations. A powerful example is the Third U.N. Conference on the Law of the Sea (UNCLOS III). The transformation of the United Nations from an instrument of the great powers into a forum for the numerous new states that have come into existence since the 1950s has created an unexpected prolifera- tion of regional and continental power centers. It has been suggested that this change owes its origin to the rapid process of decolonization or to the Cold War (see Tucker, 1977). Decolonization spawned a number of new states that could not be accommodated

within the international system without perturbation,
and the Cold War endowed the new states of the Third
World with political leverage that they could not
otherwise have assumed.[2]

It is extraordinary that the claims of the Third
World have elicited an unusually favorable response
from the liberal elites of the industrialized
countries whose concern with domestic inequalities in
income and wealth has paralleled their growing
sensitivity to international inequalities. The pro-
ponents of the new egalitarianism in the weak states
find themselves unexpectedly blessed with powerful
sympathizers in the West. The motives of these
liberal elites, however, appear to be different from
those of the political leaders of the South. While
Western intellectuals associate the new egalitarianism
with the greater equality of individuals within the
domestic environment and the evolution of a supranational regime in the international arena, the
claimants of the South associate the new egalitarianism with the equality of states rather than
individuals and with increasing nationalism and
jealously guarded sovereignty in the international
society. Indeed, it would appear that inequalities
among individuals in the new states have increased
rather than lessened. Autocratic regimes abound in
the developing nations and the distribution of wealth
is perniciously asymmetric.

For the West, interdependence implies some
sacrifice of sovereignty, while for the South interdependence is seen as a means of extracting maximum
concessions from the nations that possess a majority
of the globe's wealth and power. The Western liberal
elites see an opportunity for growing supranationalism
in the new egalitarianism, while the Southern elites
see an opportunity for growing nationalism. The
dichotomy of motives and the moral imperatives
behind these motives works resoundingly in favor of
the Third World challengers.[3]

The new egalitarianism maintains that any
international economic order that seeks to prom
the principle of equality of opportunity yet in s
striving fails to discriminate in favor of the
disadvantaged is itself discriminatory and unjust
(see Tucker, 1977, p. 68). This seems to be an
articulation on an international level of the
Rawlsian theory of justice, an articulation that
Rawls himself confines to the level of domestic
society (Rawls, 1971).

The claims of the Southern elites, when shorn of
ideological posturing, are seen to be a blatant play
for the redistribution of power and wealth, an
adjustment in the world order that is to be achieved
through states for purposes elucidated by states
alone.

Interdependence among nations implies reciprocal
effect. The ability of countries to influence each
other, however, is usually uneven, and this leads to
unequal degrees of dependence. This assymmetry can
be a source of power since the less-dependent nation
can often use it to manipulate the links that bind
(see Nye, 1978, p. 132).

The degree to which one nation can influence
another depends on how "sensitive" or "vulnerable"
each nation is. Here a nation is "sensitive" if it
is liable to penalties imposed from the outside but
can, with a change of policies, minimize or eliminate
those penalties. A state is "vulnerable" if it is
liable to expensive effects imposed from the outside
even after efforts have been made to modify or escape
the situation. Both sensitivity and vulnerability
are matters of degree and are functions of the time
and expense required to develop alternatives (see
Ney, 1978).

Interdependence forces nations into postures of
both conflict and cooperation as they exploit or are
exploited by the situation. In relation to weaker
countries interdependence is encouraged as promoting

while in relation to stronger
ship is often decried as an
ir concessions.[4] The dynamics
such that in most cases the
solution is preferable to the
ict outcome. Cooperation is
t on sovereign policy.
inherent in interdependence
_____on to the pressures toward
egalitarianism, that mitigates against unilateral
exploitation of ocean bed resources by the West. It
is held, for example, that the threat of producer
cartels will discourage the advanced industrialized
countries from offending the rest of the world too
much. There is a real possibility of short-term
price and supply disruptions in several resource
markets if the Third World pursues policies of
cartelization (see Pindyck, 1976). There is,
however, considerable dispute over how efficient
these cartels might be in the long run. (If willingness to use buffer stocks can be presumed in cases of
economic crises, then, in fact, net consumers are
most likely able to insulate themselves and withstand
greater shocks.)

Several natural resource markets are characterized
by factors conducive to the formation of temporary
cartels. These factors are inelastic demand, concentrated production, high market share of potential
cartel members, and potential members having a cost
advantage over fringe suppliers.

Cartels and commodity stabilization schemes that
are weak de facto cartels already exist or are being
planned for crude oil, industrial and gem diamonds,
sugar, copper, cotton, coffee, rubber, cocoa, tin,
tea, jute, sisal, and bauxite. Depending on the
success of these experiments, other markets may also
be cartelized. Cobalt, for instance, is cartelized
for all practical purposes, with Zaire acting as a
monopolist with a competitive fringe.

The primary source of cartel breakdown
pressure from the competitive fringe. In t
a possible cobalt, copper, or nickel cartel, ᴛₕₑ
supply pressure could come from deep ocean minerals.
    The exposure of the U.S. economy to disruptions
in foreign sources of supply has reached a stage
where the U.S. government has begun to give the
matter its serious attention. The creation of a
cabinet-level Mineral Review Committee is testimony
to this concern.
    It may be observed that the current U.S. supply
of cobalt and manganese, both obtainable from deep
ocean mining, is almost entirely imported. The raw
materials import position is worse for Western Europe
and Japan. While the United States imports 15% of
its total consumption of industrial materials,
Western Europe imports 75% and Japan 90%. The
vulnerability of its principal trading partners
indirectly exposes the United States to economic
instability since disturbances in one economy might
be transmitted to another through the myriad links
of international trade, investment, and finance that
bind them (see Mining and Materials Policy, 1976,
p. 108).
    Those who take the threat of Third World resource
cartels seriously argue that in addition to restrict-
ing supplies, the countries of the South can increase
their leverage in the international arena by capital-
izing on disputes in the United States/Europe/Japan
economic triangle just as in the past they have
exploited disputes with the United States/USSR/China
political triangle.[5] Others take the position that
there are strong factors weighing against the
possibility of producers' cartels modeled on OPEC.
The alternatives of stockpiling, substitution, and
recycling would ultimately create enough supply
pressures to disintegrate most resource cartels.
Even those who take this more optimistic stance do
concede, however, that there probably are

opportunities for producer alliances for a few
nonfuel minerals.

There are other arguments for reducing material
inequalities and, by implication, sharing the
benefits to be derived from the exploitation of ocean
bed minerals. Not only could the South use its links
of interdependence to transmit instabilities to the
West by fostering rivalries among great powers,[6] but
they could take direct action through economic or
military war as domestic pressures become uncontain-
able (see Tucker, 1977, pp. 84085). This scenario
seems a little extreme, however, since the developing
nations themselves are driven by differences that
could preclude unified action against the West.
Moreover, they do not possess military institutions
or instruments strong enough to conduct a campaign
of any significance, without help from major foreign
sources.

AN UNUSUAL INDUSTRY PROBLEM
The strategic responses of industry, within the
Western capitalistic system, to the promises and
threats posed by the new resource will largely
determine the structure of the emergent ocean mining
industry. The promises of ocean mining are similar
to those of any investment in minerals: an opportunity to make a reasonable profit, and a means of
assuring a stable supply of cobalt, copper, nickel,
and perhaps manganese from a source less vulnerable
to disruption than are many current supplier nations.
The threats, however, are unusual and complex, and
the strategic response to those threats may well
dominate the evolution of the industrial organizational form that will eventually be responsible for
the commercial development of ferromanganese
deposits.

The threats arise from the uncertainty that seems
pervasive in the operating environment of the
fledgling industry. The four main areas of uncertainty are scientific, legal, technological, and

economic. Each of these uncertainties has
implications for efficient resource management and
for the industry response.

The design of resource management policies is
rendered difficult by the considerable scientific
uncertainty that attends the reserve base of ferromanganese nodules. The uncertainty is made up of two
parts. The first part is the familiar question about
the shape of the supply curve, and to this extent
ferromanganese nodules are no different from any
other mineral deposit. The second part is unique to
these deposits: important questions still remain
regarding the evolution, rate of formation, and
composition of the nodules, leaving the abundance and
grade of the resource on the seabed open to controversy and uncertainty.

Specific policies with respect to optimal
exploration and depletion are thus almost impossible
to design. Yet the need for them has never been
greater, both because sound global resource management strategies are becoming increasingly important
and because the economic and management approach to
ocean mining will set important precedents for the
development of other supranational resources, such as
those of the Antarctic, asteroids, and space.

The legal uncertainty arises from the transitional
status of the resource. First, there is some question
whether the resource will, in fact, be transformed
from a transnational into a supranational one.
Second, there is considerable confusion about the
operating rules that will emerge following a successful transformation. (If the transformation does not
take place, presumably the United States will pursue
a policy of unilateral legislation, imposing its own
operating rules on industry.)

In addition, the deep seabed mining bill recently
passed in the House places specific limitations on
the type of treaty the United States would ultimately
consider. According to the bill, the treaty at a
minimum must provide U.S. citizens assured and

nondiscriminatory access to deep seabed sites and not
impose any restriction that would threaten investments already made. This draft U.S. legislation is
already serving as a model for similar proposals in
other countries. Eventually a web of ad hoc,
informal, reciprocal agreements between the U.S.,
West European, Japanese, and Canadian governments
could result.

Such a de facto multinational deep seabed mining
regime, outside the purview of the Law of the Sea
negotiations, could destroy the chances of a truly
international treaty encompassing the desires of all
150 nations taking part in the UNCLOS III
negotiations.

If the transformation does succeed, then the Law
of the Sea negotiating text, as presently written,
would create a Seabed Authority to operate the
"Enterprise," a mining arm of the Authority; render
corporate access to mine sites conditional on the
transfer of technology to the Enterprise and on
revenue sharing; and set production limits to prevent
the destabilization of the international markets for
cobalt, copper, manganese, and nickel. These rules,
however, are merely the most recent in a series that
has evolved over time. Their acceptance is by no
means assured. The uncertainty the industry feels
about the final outcome is justified in view of the
range and scope of schemes that have been proposed
in the past five years. The spectrum of these
schemes, from passive to active, are as follows:

1. Registry: A Seabed Authority serves as an
international registration and clearing office so
that claims can be filed and registered by mining
firms or international consortia. This would
ensure the exclusivity of claims. The Authority
collects rent in the form of registration fees.

2. Licensing: The role of the Authority is
broadened by giving it the power to license firms.

This would increase the scope of the regulatory
activity.

3. Quota Licensing: The Authority can restrict
the number of mine sites allocated to any one
country.

4. Banking: An area's mining firms must submit two
mine sites to the Authority for licensing. The
Authority licenses one site to the firm and keeps
the dual site in a site bank. Over time, a number of
sites would accumulate and could be made available to
developing countries that do not presently possess
the capital or technology to undertake deep ocean
mining.

5. Permanent Mixed: The dual mine site is exploited
directly by an arm of the Authority known as the
Enterprise.

6. Phased Mixed: The mining firms are phased out
over time so that ultimately only the Enterprise
mines the manganese nodules.

7. Joint Ventures: The ocean mining firms participate only as minority joint-venture partners with
the Enterprise.

8. Enterprise Only: From the outset, the Enterprise
has a monopoly on deep ocean mining.

    Associated with each institutional arrangement,
it has been suggested, could be a system of production controls arrayed along a spectrum:

1. No production controls.

2. Land-based mineral producers to be compensated
for loss of revenue by transfer payments.

3. Stabilization of commodity prices through buffer
stocks maintained by the Authority.

4. Direct integration of the Authority's activities
with commodity price stabilization schemes.

5. Production quotas assigned to ocean mining. One
suggestion has been to restrict ocean mining to a
level that corresponds to a 6% annual increase in the
world demand for nickel.

The technical uncertainty relates to the use of
technologies that remain to be proven. Only a few
tentative pilot operations have been undertaken, but
whether the results obtained can be directly trans-
lated into parameters that govern large commercial
operations is not obvious. There are technological
uncertainties in almost all aspects of the mining and
processing industry.

The development of ferromanganese nodules
involves several distinct operations, each with its
own technical requirements. The nodules have to be
harvested from the ocean floor, which is not easy
because of the unevenness of the deep seabed and
differences in sediment consistency. They must then
be raised to the harvesting ship and probably trans-
ported to a land base for processing. (Some processing may be undertaken on board the harvesting
ship itself, depending on relative energy costs.)
Finally, the waste products from the processing
operation must be disposed of on land.

The most widely considered mining and recovery
system consists of an ocean surface platform, a
conveyor pipe to transport the nodules through the
water column, and a gathering device to collect the
nodules on the ocean floor. There are several
different engineering designs possible for each
component.

The gathering device could take several configurations. Some that have been suggested are:

An Unusual Resource Management Problem

1.   The applied Nodule Collector, which is a square
U-shaped frame mounted on a sled and dragged along
the ocean floor. This has been designed by Deep Sea
Ventures, Inc., a U.S. corporation (now Ocean Mining
Associates).

2.   The Drag Net, which is like a steel herring trawl
having meshes that gradually reduce in gage from the
mouth to the tapered end.

3.   The Activated Collecting Device, which has been
constructed by DEMAG AG, a West German firm. The
device consists of a case containing control equipment, lights, and a television camera. It is mounted
on a tracked undercarriage. The case has a mobile,
cantilevered suction pipe in the front with a rotary
dredge head that can be moored vertically.

4.   The Wide Pick-up Device, which sets up large-
scale tract dumps at intervals of 300 to 400 meters.
The collecting device is moored by means of a hose
and travels to and fro between two side machines,
where it climbs a ramp and discharges the nodules.

   The vertical transportation of the nodules to the
surface barge could be carried out in a number of
ways, three of which are:

1.   The Continuous Line Bucket System, which uses
buckets attached to a continuous cable and transports
the nodules from the floor to the barge.

2.   The Airlift System, which forces compressed air
into a vertical pipe, thus forming a mixture of
water, sediment, and nodules. This mixture is then
forced up the pipe by means of a device called a
"mammut pump."

3.   The Hydraulic System which uses a pump installed
on a vertical pipe to force the nodules up. This

'ficient than the pneumatic system
f the pump is mounted in a water
)0 meters.

₋₋ ₙave been pumped up, they must be
₋₋ₐ𝖼ed from water as well as sediment. This must
be done on a surface barge which, again, could take
several configurations, of which three are:

1. A conventional ship with a central trunk.

2. An operational base, similar in form to a ship.

3. A semisubmersible.

→ The economic uncertainties involve capital and
operating costs and the future price trajectories of
cobalt, copper, manganese, and nickel. The price
uncertainties are due to the changing structure of
the current world markets for these minerals and the
unknown impact of ocean-based production on those
markets.
     A major source of economic uncertainty is the
processing plant, which will account for at least
50% of total cost. Compounding the engineering
problems associated with the plant are locational
issues. The expenses relating to plant location may
well be considerably higher than initially anticipated. The optimal location for a plant is one where
there is a large body of water and exceptional access
to transportation systems. This would suggest
locating a plant on the coastal zone of the western
states of the United States, property which is
extraordinarily expensive. Moreover, extremely
restrictive land-use regulations and the major
environmental issue of tailings disposal would make
it very difficult to establish a plant in the desired
area. Even if initial approval to locate a plant is
obtained, the normal regulatory lag and the mechanics
of securing the necessary permits may add considerably
to costs.

⟶ The changes in the structure of world markets for
cobalt, copper, and nickel that are currently occurring also have implications for the future price
trajectories. But the impact of these changes has
been insufficiently analyzed so far to allow confi-
dent prediction of the competitive shape of the
industry in 10 or 15 years. Disputes over land-based
supplies of manganese and uncertainties about the
size of demand add to the problem.
⟶ Overall, there is a tendency for decreasing
concentration among the noncommunist producing
countries. In cobalt, however, the production is
still largely in the hands of Zaire and Zambia, and
together the two form an oligopoly that effectively
sets world prices, with leadership mostly maintained
by Zaire. In times of curtailed production in Zaire,
as at present, Zambia has shown no hesitation in
setting world prices by exploiting its market power.

In the world copper market, the pressure by
developing countries to expand output despite low
prices is a reflection of both their low marginal
costs of production and their objective of maximizing
foreign exchange and employment.

The most significant change is that in the nickel
market, since ocean mining must be viewed primarily
as a nickel venture. For many years the world nickel
market, like the cobalt market, had been characterized by a monopolistic producer and a price-following
fringe. The monopolist, INCO Ltd. of Canada, had
pretty much developed both the industry and the
markets, and its price-setting role was undisputed.
That role, however, has been seriously challenged in
recent years; and in 1977 INCO suspended its policy
of posting prices and declared that it would enter
into price competition, letting prices fluctuate
according to market conditions. This was merely a
reflection of decaying market power: in 1950 the
firm accounted for 90% of the noncommunist market;
by 1970 its share was reduced to 50% of a considerably
bigger market. INCO's comparative advantage of
control over the high-grade ores of Ontario had been

gradually negated by technological developments that
made possible the profitable production of nickel
from lateritic ores common in the developing nations.
Supply pressure at present comes from such nations
as New Caledonia, Indonesia, and Guatemala. Indeed,
overcapacity and rapidly increasing supplies in the
world nickel market have forced INCO Societe
Metallurgique Le Nickel (owned jointly by Societe
Metal and Societe National Elf Aquitaine) and Falcon-
bridge Nickel Mines Ltd. of Canada, among other
producers, to institute production cutbacks in an
effort to restore prices. Moreover, INCO, in an
attempt to control both land-based and offshore
supplies, has formed a mining consortium to develop
ferromanganese nodules. In this regard it is merely
acting like a well-behaved monopolist.

Trends in the noncommunist producer concentration
of the three minerals are shown in Table 1. The
Herfindahl-Hirschman index of concentration has been
computed for 1965, 1970, and 1975. The decline of
INCO is evident from the falling index for nickel.
The index for cobalt shows mainly the relative
changes in Zaire's preeminence and continues to be
high. The falling index for copper is testimony to
increasing global competition and the relative
decline of the United States and Canada as developing
countries increase output.

Yet another source of concern to industry is the
likely impact of ocean mining on world markets. One
recent study suggests that if ocean mining becomes
significant, there will be a substantial linking of
the world markets for cobalt, nickel, and copper and
some convergence in their prices. This development
may result in part from the fact that the relative
abundance of the three minerals in ferromanganese
deposits is very different from the relative sizes
of their world markets.

The production problem that industry faces in
deep ocean mining is one of optimizing joint returns
from cobalt, nickel, and copper as co-products.

Table 1
Herfindahl-Hirschman Index of Noncommunist
Producer Concentration

	Cobalt	Nickel	Copper
1965	0.349	0.590	0.172
1970	0.492	0.385	0.167
1975	0.394	0.231	0.131

Consequently, the real price of cobalt may fall to
almost the marginal cost of production from ocean
mining in order to increase demand substantially. As
presently delineated, two mine sites could easily
satisfy the projected 1985-1990 demand for the metal.
In order to meet the constraints of co-production and
prevent gross underutilization of capacity, demand
for cobalt will need to be stimulated greatly through
a lower price. The joint-profit-maximizing condition
will also call for a rise in the real price of nickel
from current, rather than depressed, levels.

    Much of this uncertainty can be translated into a
high variance in the stream of income that would flow
from an ocean mining investment. This high variance,
in turn, implies high risk. To the extent that firms
will view an ocean mining venture as one element in a
portfolio of investments, it will have to have either
a very low correlation with the other ventures of the
firms (so as to reduce the systematic risk of the
entire portfolio) or a return high enough to put it
on the efficient frontier for the investing firms.
The search for a high return and/or diversification
of risk will influence the corporate response to
ocean mining.

    The industry response to the uncertainty and risk
has been to recognize that the traditional multinational corporation (MNC) form of industry in the
three (or possibly four) hard minerals must evolve
into a Muntinational Resource Development Consortium
(MRDC) industry. The MRDC brings together mineral-

producing companies, suppliers of capital, developers
of ocean mining technology, and users of minerals.

The incipient ocean mining industry is characterized by a number of consortia composed of firms and
organizations drawn from the nations of the Organization for Economic Cooperation and Development (OECD).
Each consortium consists of a number of firms clus-
tered around a dominant firm that is the chief
repository of the technology needed to develop ocean
minerals. The other firms provide knowledge of
markets, metal-refining expertise, capital, and, in
some cases, specialized knowledge of transportation
systems. Some of the participants are present producers of the metals, such as International Nickel
and Union Miniere, who are concerned about securing
additional reserves. Others are motivated by the
desire to diversify their operations (e.g., Kennecott
Copper; Rio-Tinto-Zinc) or an interest in exploiting
a special technological advantage (Lockheed Aircraft
Corporation; CNEXO, the Centre National pour
l'Exploitation des Oceans), and a few others are oil
corporations (Standard Oil of Indiana; British
Petroleum) considering expanding into nonfuel
minerals.

Four of the consortia are based in the United
States, while the fifth is based in France. The four
U.S.-based consortia are privately owned, while the
French consortium is a joint venture involving three
public organizations and two private firms. The
countries from which the participating firms origi-
nate are the United States, Canada, France, Belgium,
the Netherlands, West Germany, Japan, and the United
Kingdom.

The MRDC creates an integrated corporation by
combining components from different corporations,
and the strategy of the MRDC must reflect the objectives of all constituent companies rather than that
of a single dominant parent. This creates unique
strategic planning and management problems and
opportunities. The superficial similarity with an

international oil consortium is that often there is a single large shareholder associated with a number of smaller shareholders. While in the oil consortium the largest shareholder is typically the largest contributor of capital, in the MRDC it is the firm that possesses the technology needed to develop ocean minerals. The supplier of technology has the potential to dominate the MRDC.

The MRDC and MNC are elements of an evolutionary chain. They are linked by a corporate configuration devised to manage another international resource, the radiofrequency spectrum. In both cases the development of a technology that made possible the exploitation of a known resource spurred the search for an appropriate organizational framework for that exploitation. The chain of events that led to the formation of INTELSAT are part of a process that started several years ago and is continuing even now. The process giving rise to MRDC, however, differs in several ways:

1. The debate over the utilization of the ocean bed resource is not between the United States and Western Europe, as it was in the case of communication frequencies, but between the netproducers and the netconsumers of the resource.

2. INTELSAT, unlike the MRDCs, was not competing with existing producers of products.

3. The response has been the creation of four private consortia and possibly one intragovernmental arrangement. The communication problem was solved by creating a single corporation.

4. In the case of INTELSAT only one country had the technology and only one company had the authority to manage this technology.

5. The development of communication frequencies

required the cooperation of several nations.
Unilateral development could not have been a solution,
whereas with ocean mining it is quite efficient for
the country that possesses the technology to undertake unilateral development.

In the case of INTELSAT, the U.S. corporation
involved and the U.S. government tried to pursue a
policy of keeping the technology and sharing only
the rents of development. The West Europeans, whose
cooperation was vital, were dissatisfied with this
arrangement and used their bargaining power to
acquire a share of the communications technology.
This was done through subcontracting arrangements
that deemed that a certain percentage of the work be
done by European firms.
The West Europeans agreed to make only a
temporary arrangement and demanded that the decision-
making power be altered at the end of five years to
better reflect the changed relative bargaining positions and the generational changes in the technology.
The arrangement of an exclusive intergovernmental
consortium subcontracting to industry was altered at
the end of five years to permit the formation of
several consortia. Hence, the model for the new
economic order was really provided by West Europeans
bargaining with the United States for an equilizing
(sharing) of technological assets. The West Europeans had the infrastructure to absorb the U.S.
technology and, indeed, could have developed it
themselves. In the case of the debate between the
North and the South, the latter do not have an
infrastructure on which to build the technological
systems needed for ocean mining, even if a transfer
of technology took place. The South has very little
long-run technological leverage over the North.
Besides the MRDC, there are possibly two other
evolutionary organizational forms that industry could
take: the Enterprise and a Group of 77 Corporation
(named after the developing nations club, the Group

of 77). The Enterprise would be the first supranational corporation, created by an international
treaty and reporting to a Seabed Authority. As the
operating arm of the Seabed Authority, the Enterprise,
at least for the first two decades of ocean mining,
would need to buy its technology and services from
industry.

A Group of 77 Corporation could be a commercial
ocean mining activity managed by professionals from
countries such as Brazil, India, and Nigeria, funded
by OPEC, and obtaining its technology under a
licensing agreement from a developer such as Lockheed
Corporation.

NOTES

1. Joseph S. Nye comments, "As military force has
become more costly for large powers to apply, power
has become less fungible and the traditional hier-
archy of states has been weakened. There is an
increased discrepancy between power measured in
military resources and power measured in terms of
control over the outcome of events" (Foreign Policy,
spring 1976, p. 142).

2. Daniel Bell is of the view that two major
structural transformations contributing to international instability are the transition of Western
industrial countries into more open and egalatarian
societies and the "emergence of a bewilderingly
large number of new states of vastly diverse size,
heterogeneity, and unevenly distributed resources.
As a result of this development, the problem of
international stability in the next 20 years will be
the most difficult challenge for those responsible
for the world policy" (Foreign Policy, summer 1977,
p. 111).

3. Zbigniew Brzezinski has stated that "the problem
of the less developed nations is the moral problem of

our time....Intensified social strife and global
animosity are bound to be the consequence of mankind's
failure to tackle the problem of global inequality"
(Foreign Affairs, July 1973, p. 276).

4. In the case of one developing country, Leo E.
Rose writes that "interdependence based on inequality
among the nations of South Asia is acceptable to
India; on a global level, however, interdependence is
denounced as a conspiracy by the developed nations to
maintain the status quo" (In J. Rosenau, K. W.
Thompson, and G. Boyd, eds., World Politics, Free
Press, 1976, p. 207).

5. C. Fred Bergsten believes that "a wide range of
Third World countries...have sizeable potential for
strategic market power....Supply countries could
exercise maximum leverage through withholding
supplies altogether, at least from a single customer
such as the United States" (Foreign Policy, summer
1973, p. 108).

6. Helmut Schmidt has suggested that the South will
have to be appeased somehow in order to secure international stability. He writes that "in the long run,
there will have to be more genuine transfers of real
resources in order to provide the less-developed
nations with a genuine basis for continued self-
development and thus also to decrease social and
political tension" (Foreign Affairs, April 1976,
p. 450).

BIBLIOGRAPHY

Aron, Raymond, 1968. Progress and Disillusion: The
Dialectics of Modern Society, Praeger, New York, NY.

Bell, Daniel, 1977. "The Future World Disorder,"
Foreign Policy, no. 27, summer 1977.

Bergsten, C. Fred, 1973. "The Threat From the Third World," Foreign Policy, no. 11, summer 1973.

Brzezinski, Zbigniew, 1973. "U.S. Foreign Policy: The Search for Focus," Foreign Affairs, no. 51, July 1973.

Business Week, May 9, 1977, pp. 67, 81.

Erb and Kallab, 1976. Beyond Dependency, Overseas Development Council, Washington, D.C., May 1976.

Hopkins, Raymond, 1976. "International Role of 'Domestic' Bureaucracy," International Organization, vol. 30, no. 3, summer 1976.

Kildow, J.T., et al, 1976. Assessment of Economic and Regulatory Conditions Affecting Ocean Minerals Resource Development, Report to the Department of Interior, 1976.

Mining and Materials Policy, 1976. Annual Report of the Secretary of the Interior to the U.S. Congress.

Nye, Joseph S., 1976. "Independence and Inter-dependence," Foreign Policy, no. 22, spring 1976.

Pindyck, R.S., 1976. "Gains to Producers from the Cartelization of Exhaustable Resources," World Oil Project Working Paper, MIT EL-76-012-WP, May 1976, Cambridge, MA.

Rawls, John, 1971. A Theory of Justice, Harvard University Press, Cambridge, MA.

Schmidt, Helmut, 1976. "The Struggle for the World Product," Foreign Affairs, vol. 52, no. 3, April 1976.

Tucker, Robert W., 1977. Inequality of Nations, Basic Books, New York, NY.

PART II

THE VALUE AND ABUNDANCE OF THE RESOURCE

RESOURCES IN SEAFLOOR MANGANESE NODULES

Jane Z. Frazer

Various authors have estimated tonnages of seafloor managanese nodules and the amounts of metals they contain or the number of available mine sites. These estimates are generally assumed to be answers to the question, "What is the magnitude of the nodule resource?" When we compare the estimates, there seems to be utter disagreement. Predictions range from Mero's (1956) estimate of more than 1.5 trillion tonnes of nodules in the Pacific Ocean containing 16.4 billion tonnes of nickel, to a paper by Bastien-Thiry et al. (1977) which concludes that there are 670 million tonnes of nodules with a reserve of 6 million tonnes of nickel.

When we look into the reasons for these enormous discrepancies, more than three orders of magnitude, we find that the predictions are not wildly different answers to the same question but answers to different questions. Mero was estimating total nodule tonnages that might occur in the Pacific Ocean, and he stated clearly that the amount that could be economically mined would be perhaps 10% or even only 1% of that total. The French group, on the other hand, was considering (1) only the area in the northeastern Equatorial Pacific that they had explored, not the whole ocean; (2) only the deposits within that area which meet grade and abundance requirements for first-generation mining; and (3) not the tonnages that exist in situ but only those which could be expected to be recovered with first-generation mining and processing technology.

We can evaluate the reliability of the various predictions only if we keep in mind exactly what is being predicted. In this paper I will consider separately some different resource categories as defined by the U.S. Department of the Interior. I will review predictions that have been made so far, the amount and reliability of the data on which they

were based, and the assumptions made by each author.
I will attempt to explain the discrepancies that
exist and identify new information that is required
to produce more precise predictions.

DEFINITIONS
Figure 1 is the classification chart for mineral
resources adopted jointly by the U.S. Department
of the Interior's Bureau of Mines and the Geological
Survey in 1974. According to the USDI terminology,
mineral reserves include only those identified
deposits from which raw materials can be produced
economically under the legal regime and market conditions prevailing at the time the estimate is made.
In this strict sense, there are no seabed mineral
reserves. There seem to be no organizations willing
to begin ocean mining until a satisfactory legal
regime is established. According to most observers,
a rise in nickel prices is also required for nodule
mining to begin.
    Paramarginal resources are defined as "the portion
of subeconomic resources that (a) borders on being
economically producible or (b) is not commercially
available solely because of legal or political circumstances." In order to be considered paramarginal
the nodule deposit must meet minimum grade and
abundance (concentration) requirements.
    Under present conditions nickel is expected to
account for 70% of the metal values in manganese
nodules, with the remainder coming mainly from copper
and cobalt (Pasho, 1979). Prospective miners have
been seeking deposits high in nickel and copper, and
the required average grade is generally accepted to
be about 2.25 weight percent copper plus nickel, with
a cutoff grade of 1.8% (Archer, 1976; Kildow et al.,
1976).
    Concentration, defined in terms of kg/m^2 of
nodules at or within a few centimeters of the sediment surface, must be adequate for economic recovery
of nodules by first-generation mining equipment. A

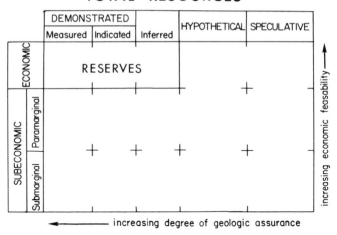

Figure 1.   U.S. Department of Interior classification system for mineral resources.

reasonable estimate for the required minimum concentration is 5 kg/m^2 (Kaufman, 1974).

Seafloor topography in the area to be mined must
be smooth enough and flat enough to permit economic
operation of first-generation mining equipment.
Unfortunately, there is so little detailed topographic
information available that most predictions must
ignore this requirement.

Submarginal resources are "the portion of subeconomic resources which would require a substantially
higher price...or a major cost reducing advance in
technology." In this category we would include nodule
deposits that do not meet the requirements specified
above but which might be minable at some time in the
future. Participants in the 1977 UN Workshop on
Seabed Mineral Resource Assessment agreed that a combined nickel and copper grade of about 1.5% would be
a reasonable lower limit for submarginal resource
grade, but there is apparently no minimum concentration that can be specified (Holser, 1976). Thus
submarginal nodule resources consist of deposits with
nickel plus copper in the range 1.5-2.25% as well as
those deposits with higher grade where concentrations
are less than 5 kg/m^2.

Independent of the level of economic value, we
can classify resources according to the extent of our
geologic knowledge about them. Demonstrated resources
include indicated resources, "material for which
esimtates of the quality and quantity have been computed partly from sample analyses and measurements
and partly from reasonable geologic projections," as
well as measured resources, "material for which estimates of the quality and quantity have been computed,
within a margin of error of less than 20%, from
analyses and measurements from closely spaced and
geologically well known sample sites."

Other resources are categorized as inferred,
"material in unexplored but identified deposits for
which estimates of the quality and size are based on
geologic evidence and projection," or undiscovered,

"surmised to exist on the basis of broad geologic knowledge and theory."

AVAILABLE NODULE DATA
Let us consider briefly the size and the reliability of the public data base, from which most nodule resource estimates are made. Most of the data are stored in the Scripps Institution of Oceanography Sediment Data Bank and derive from other oceanographic institutions and a variety of published and unpublished reports, including some industry sources. In 1972 the data bank contained fewer than 500 nodule assays, but today there are about 4000 assays from 2000 sampling locations. Additional data stored includes information about nodule concentration from seafloor photographs, grab samplers, and box cores, as well as more than 50,000 station locations for which nodules have been reported as present or not.

Figure 2 shows the locations of available nodule assays in the Pacific Ocean. More than half of the assays and almost all the concentration data are from the region between the Clarion and Clipperton Fracture Zones, indicated by the dotted lines. Assay sites in the Indian Ocean are distributed about like those in the western North Pacific, and they are even sparser in the Atlantic Ocean. In order to give some idea of the widely scattered nature of the data, I have superimposed data points for a typically sampled area of the Indian Ocean on a map of the United States, which covers an area of about the same size (figure 3).

That is the bad news. The good news is that nodule grade is rather consistent over large distances, and analyses of nodules from only a few locations usually give a reasonably good picture of the average nodule composition within an area. To illustrate this point, I have selected several areas where the amount of available data has increased dramatically within the past five years. Table 1 compares the average composition of nodules in each area from data now available with the averages for the same areas

Figure 2. Map of the Pacific Ocean showing locations
where manganese nodules have been assayed. Data from
Scripps Institution of Oceanography Sediment Data
Bank, July 1978. Dashed lines indicate the Clarion
and Clipperton Fracture Zones.

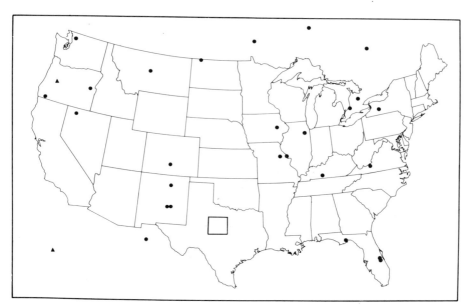

Figure 3. Nodule assays from part of the Indian
Ocean, 10-30°S, 80-110°E, an area equal in size to
the United States, superimposed on a U.S. map to
indicate the data distribution. Assays are indicated
by filled circles; seafloor concentrations have also
been measured at sites marked by filled triangles.
The box indicates 40,000 km^2, the size of a typical
mine site.

Table 1
Comparison of average nodule grades determined from large and small data bases

Area	1972 data[a]		1978 data[b]	
	No. assays	Avg. Ni+Cu	No. assays	Avg. Ni+Cu
Clarion-Clipperton Zone (7-18°N,115-155°W)	32	2.24 $\pm$ .76	1115	2.29 $\pm$ .61
Domes Site A (8-10°n,149.5-151.5°W)	3	1.54 $\pm$ .67	109	2.11 $\pm$ .70
Domes Site B (11-12.5°N,137.5-139.5°W)	4[c]	2.77 $\pm$ .61	10	2.82 $\pm$ .16
Domes Site C (13.5-16.5N,124.5-127.5°W)	4	2.28 $\pm$ .61	249	2.41 $\pm$ .37
Japanese Geological Survey Area (4-10°N,165-175°W)	8	1.85 $\pm$ .89	55	2.00 $\pm$ .63
Manihiki Plateau Area (10-20°S,160-170°W)	7	0.37 $\pm$ .23	69	0.43 $\pm$ .23

a. Data from Horn et al. (1972).
b. Data from SIO Sediment Data Bank, including data from Horn et al. (1972).
c. No assays available from this area; nearby stations used.

that could have been calculated solely from data
available in 1972 (Horn et al., 1972). Results from
the two data sets are remarkably similar, and it is
clear that few assays are required to indicate whether
a nodule deposit is likely to be within the grade
ranges for paramarginal or submarginal resources.

The public data base is frequently criticized for
inaccuracy by those who do not like the predictions
based on it. I have concluded that most of the
assays are probably correct to within 10-15% (Frazer,
1979). Additional inaccuracies are introduced by the
analysis of undried samples, nonrepresentative portions of nodules, or single nodules that do not
represent the bulk of nodules from a given location,
but these errors tend to cancel each other when assays
are averaged over an area.

A serious flaw in the public data base is the
lack of sufficient data on nodule abundance, which is
much more variable than grade over short distances.
There are problems in determining actual nodule concentration on the seafloor from grab samples or
photographs, and errors are likely to be in the
20-25% range (Bastien-Thiry et al., 1977). There are
also disagreements among some investigators about how
to interpret results from grab samplers and photographs, so there may be some systematic differences
between concentrations reported by different
researchers. Box cores are the best device for
determining nodule concentration; but they are expensive to collect, and not many have been collected
outside of the Clarion-Clipperton Zone.

In order to make accurate estimates of nodule
resources, we need many more measurements of nodule
concentration. We also need to develop a standardized
method for interpreting such measurements.

DEMONSTRATED PARAMARGINAL RESOURCES
As Holser (1976) pointed out in a special note at the
end of his paper, only the deposits between the
Clarion and Clipperton Fracture Zones in the north-

eastern Equatorial Pacific have been well enough
explored to be considered demonstrated resources.
This area, roughly 7-18°N and 115-155°W, covers about
6 million km^2 (see Figure 2). Using data compiled
originally by Mero (1965), Spangler (1970) was the
first to suggest this region as the best for commercial exploitation of manganese modules. Since many
authors fail to separate the Clarion-Clipperton Zone
from the surrounding area, we will include economically minable deposits from this whole northeastern
equatorial region (about 0-30°N, 170°E to 100°W) in
the demonstrated paramarginal resource category.

Resource magnitude is frequently expressed in
terms of the number of mine sites available for
exploitation immediately or in the near future. In
order to be considered a mine site, a nodule deposit
must contain sufficient reserves to meet minimum
economic production requirements; the amount depends
on our assumptions about these production needs and
about mining efficiency (the portion of the mine site
that is likely to be accessible to mining equipment,
the portion of in situ nodules that can be recovered).
Since there is considerable disagreement on these
matters, resource estimates in this paper will be
expressed solely in terms of tonnages of economically
minable nodules as defined above (grade greater than
or equal to 1.8% nickel plus copper with an average
grade of 2.25%, concentration greater than or equal
to 5 kg/m^2) and the amounts of metals they might
contain (table 2). These tonnages are often buried
within the papers, not listed as the main conclusions.
Where authors have predicted only the number of mine
sites or the amount of recoverable materials, I have
converted the results to this standard basis for comparison using the authors' own assumptions about
production needs and mine efficiency. Where authors
have selected a different grade or abundance level to
define economically minable nodules, I have estimated
from information in the paper what tonnages they
would have predicted that meet the specified criteria.

Table 2
Demonstrated paramarginal resources in the Clarion–Clipperton zone
(Amounts in situ in millions of tonnes)[a]

Source	Nodules	Nickel	Copper	Cobalt	Manganese
Archer (1975)	12,400	106	87	87	2,400
Holser (1976)	14,000	120	98	24	2,700
Pasho and McIntosh (1976)	11,900 (1,600–50,600)	38 (14–352)	79 (9–324)		
Pasho (1977)	(3,600–34,200)				
Frazer (1977)	13,800 (6,900–27,600)	121	100	23	2,400
Mero (1977)	54,000	650	520	115	11,000
Bastien-Thiry et al. (1977)	4,750 (3,900–5,600)	41	34		
McKelvy et al. (1978)	15,000	135	104	23	2,600
Frazer (this work)	(4,000–15,000)				

a. Each author's results were converted to the common basis of in situ tonnes.

Mero's 1965 estimate of nodule resources was based
on only 54 nodule assays, 29 seafloor photographs, 10
grab samples, and 62 cores for the entire Pacific
Ocean. His observations about where nodules are
likely to be abundant and where they are enriched in
certain elements were remarkably accurate considering
the small amount of data he had. Nevertheless,
although his early resource prediction is still frequently quoted, it should now be set aside in favor
of more recent estimates based on a much larger base.

Archer (1976) studied the prospects for
exploitation of manganese nodules with a data base
that had been compiled during the first phase of the
IDOE Managanese Nodule Program (Ewing et al., 1971;
Horn et al., 1972; Frazer and Arrhenius, 1972). He
had available nodule assays for 530 locations, 48 of
them within the Clarion-Clipperton Zone, and data
about the proportion of the seafloor covered by
nodules from 2310 Lamont-Doherty Geological Observa-
tory camera stations plus 146 additional camera
stations in the Clarion-Clipperton Zone from Schultze-
Westrum (1973).

Although Archer's resource estimates are for the
entire world, one can use his data on the "prime
areas" to separate out his estimate for the Clarion-
Clipperton Zone as 12.4 billion tonnes. (Tonnes =
metric tons. Tonnages of nodules cited in this paper
are in situ amounts and contain about 30% water.
Metal amounts, of course, are computed on a dry-weight
basis.) He assumed that minimum concentrations of
10 kg/m^2 would be required for seabed nodule mining,
whereas most authors now agree that 5 kg/m^2 is
sufficient. He partially compensated for this error,
however, by his estimate of 18.3 kg/m^2 for average
nodule concentration in the minable part of the zone.
This value is shown by later studies of additional
data to be much too high (Frazer, 1977; Bastien-Thiry
et al., 1977; McKelvey et al., 1978). Even the mine
site area claimed by Deepsea Ventures, Inc., which is
probably an area of above-average abundance, has only

9.7 kg/m^2 according to the "Notice of Discovery and
Claim" filed by Deepsea Ventures in 1974.

   In a rebuttal to Archer's work, Holser (1976)
also estimated the so-called potential nodule
reserves using essentially the same public data base.
Although he criticized Archer for overestimating the
required minimum abundance and underestimating expected mining efficiency, the amount of minable nodules
he seems to have predicted for the Clarion-Clipperton
Zone is 14 billion tonnes, not much higher than
Archer had estimated for this area. (Holser's
resource estimate for the Clarion-Clipperton Zone is
not explicitly stated, but from his figure 4 it can
be deduced that it is about one-third of his total
prime area estimate.)

   Later in 1976 Pasho and McIntosh presented
preliminary results on recoverable nickel and copper
from manganese nodules in the northeastern Equatorial
Pacific. Their area of study was considerably larger
than the Clarion-Clipperton Zone, including the region
between 0-30°N, 100°W - 170°E. Their analysis was
based on publicly available data including over 2000
bottom sample descriptions and appropriate nodule
analyses from the area. In addition, personal judgments were solicited from knowledgeable individuals
actively involved in industrial exploration and
exploitation efforts. These authors used a Monte
Carlo statistical technique, which combined probability distributions in the various parameters.
Their method is superior to those used by most other
authors because it results in not only an estimate of
the resource but also a statistical measure of the
error.

   Converted from recoverable tonnes to in situ
tonnes of nodules in minable deposits, Pasho and
McIntosh's estimate is between 1.6 and 51 billion
tonnes at the 90% confidence level, and about 12
billion tonnes at the 50% confidence level. (For
this conversion, I used a factor of .24 for net
mining efficiency. This is Pasho's estimate of the

50% probability level for this variable.) They report
that the most probable value is closer to 1.6 than
12 billion tonnes. Refinement of this preliminary
estimate indicated a probable range of about 3.6 to
34 billion tonnes in the northeastern Equatorial
Pacific (Pasho, 1977).

In 1977 I used a very simple grid estimator
approach to predict paramarginal nodule resources.
I divided the world ocean by an arbitrary grid at
five-degree intervals of latitude and longitude. I
then searched for prime areas by considering the only
two parameters for which significant amounts of
information existed on a world scale: nodule grade
(weighted to account for the fact that nickel is a
more valuable metal than copper) and frequency of
nodule occurrence. Any five-degree square that met
the grade and frequency criteria I had established
was considered part of the prime area. Measurements
of nodule concentration from grab samples and photographs in the northeastern Equatorial Pacific were
used to estimate the fraction of the prime area that
was likely to contain nodule concentrations greater
than 5 kg/m^2 and the average concentration for that
fraction.

My study utilized data then stored in the SIO
Sediment Data Bank, including 3100 nodule assays
from about 1500 sampling sites. My estimate for in
situ nodule tonnage in the northeastern Equatorial
Pacific in economically minable deposits was 13.8
billion tonnes, with a factor of two as the estimated
error; thus my estimate was a range of 6.9 to 28
billion tonnes. In spite of a much larger data base
and a different method, my estimate for the Clarion-
Clipperton Zone was just about the same as those of
previous authors.

Although I did not then calculate the amount of
nickel and other metals in these nodules, I recently
did so using average nodule composition for the
minable area from data now stored in the data bank;
the results are included in table 2.

Writing in 1977 on economic aspects of nodule
mining, Mero predicted an average concentration of
9 mg/m^2 over the 6 million km^2 within the North
Pacific high-grade area, a resource of about 54
billion tonnes of wet nodules containing 650 million
tonnes of nickel and the amounts of other metals
shown in table 2. Mero's average metal values for
nodules in this area are much higher than those cited
by other authors. In converting to a dry-weight
basis the results of x-ray fluorescence spectrometry
he had performed on undried nodules, he made a severe
overcorrection. Because water is relatively trans-
parent to x-rays, there is generally only a 2-3%
difference in x-ray analyses of dried and undried
nodules, not the effect of about 20% as he assumed.
Mero also assumes that the entire Clarion-Clipperton
Zone is economically minable, so he apparently did
not take the minimum grade and abundance requirements
into account as did the other authors.

The most recent prediction of nodule resources,
by McKelvey et al. (1978), uses a slightly later
version of the same public information from the SIO
data bank. They estimate that the Clarion-Clipperton
Zone has 15 billion tonnes of economically minable
nodules containing 135 million tonnes nickel and
104 tonnes copper.

Between 1970 and 1976 a French group prospected
5.5 million km^2 in the eastern Equatorial Pacific,
from 27°S to 32°N, 81°W to 170°E. Predictions based
on their exploration over an area of 2.25 million
km^2 between the Clarion and Clipperton Fracture Zones
have been reported by Bastien-Thiry et al. (1977).
Within this area they surveyed 262 localities,
sampling at 1844 stations in a nearly regular grid
pattern. At each station free-fall samplers collected
nodules and sediment, and seafloor photographs were
taken. Box cores were collected at some stations.
The report includes a brief summary of the results
but no individual data, and it does not specify the
precise location of the explored area.

The French estimate for this 2.25 million km^2
area in the Clarion-Clipperton Zone is 4.75 billion
tonnes of nodules (estimated error 18%) in situ with
the required grade and concentration. These deposits
would contain 41 million tonnes of nickel and 34
million tonnes of copper. This is less than half the
amounts predicted by most recent authors but is in
line with Pasho and McIntosh's comment that the most
probable size of the resource is closer to their 95%
confidence level prediction (1.6 billion tonnes of
nodules) than to their 50% confidence level estimate
(about 12 billion tonnes).

What factors account for the much lower resource
estimate of the French group? Figure 4 shows the
frequency distributions of nickel plus copper grade
in northeastern Equatorial Pacific nodules from the
SIO data bank (as of November 1977) and as reported
by Bastien-Thiry et al. for their exploration. Grade
distribution is not significantly different for the
two data sets, and the average values are almost
identical: 1.22% nickel, 1.03% copper, and 0.24%
cobalt for the French, and 1.25% nickel, 1.04%
copper, and 0.22% cobalt for the public data base.

The main difference between the two data sets is
in estimates of nodule concentration. From the
information available two years ago I estimated that
38% of the Clarion-Clipperton area would have concentrations greater than 5 kg/m^2 and that the average
concentration in this portion of the area would be
11 kg/m^2. McKelvey's more recent estimate was that
about 50% of the zone would contain sufficient nodule
abundance for mining, with an average concentration
of about 12 kg/m^2. The French data show average
nodule concentration in the Clarion-Clipperton Zone
to be only 3.45 kg/m^2, with 31% of the area having
concentrations above the 5 kg/m^2 minimum and that
part of the area averaging 7.3 kg/m^2.

It is difficult to dispute the French results,
since they are based on much more data from intensive
sampling over a regular grid of localities. On the

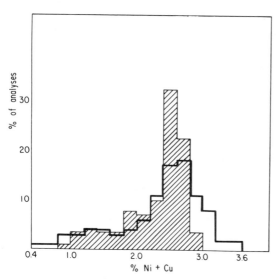

Figure 4. Frequency distributions of copper plus
nickel in the northeastern Equatorial Pacific from
the SIO Sediment Data Bank, November 1977, indicated
by the heavy black line, and as reported by
Bastien-Thiry et al. (1977), indicated by the
shaded area. (From Frazer, 1978.)

other hand, some of the French results contradict
other data which we believe to be reliable. In their
Figure 2, the French paper seems to indicate that no
deposits were found with concentrations greater than
13 kg/m^2. Kaufman (1974), however, has presented
graphs of nodule population along about 300 km of
ship's track which indicate nodule concentrations of
13-18 kg/m^2 along a significant portion of the track.
Ten percent of the concentration measurements from
the Clarion-Clipperton Zone in the SIO Sediment Data
Bank are greater than 13 kg/m^2 (figure 5). These
concentrations were determined from seafloor photographs, grab samplers, or box cores by a number of
different investigators including my colleagues at
Scripps. It is hard to believe that we are all
wrong.

Perhaps the French exploration, which covered
only half of the area between the Clarion and Clipperton Fracture Zones, did not include the areas with
the most highly concentrated nodule deposits. Or
perhaps the differences are due to different methods
for calculating nodule concentrations from
photographs.

Since the French data are unavailable, we cannot
properly evaluate their results. However, the public
data base contains insufficient information for us
to place complete confidence in abundance estimates
based on those data. Thus without additional data we
cannot determine nodule resources in the Clarion-
Clipperton Zone any more precisely than to say that
they are probably within the range 4-15 billion
tonnes. Most authors, of course, have ignored local
seafloor topography in making their estimates. We
lack the necessary information to ascertain how much
of the resource we have identified occurs in areas
that are accessible to mining equipment. This is one
good reason to place our best guess of demonstrated
paramarginal resources somewhere in the lower part of
this range.

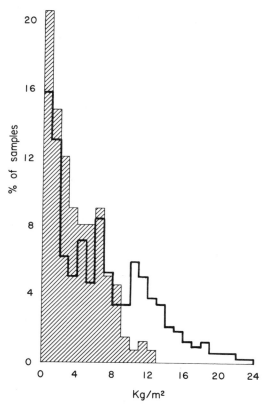

Figure 5. Frequency distributions of nodule concentration in kg/m^2 in the northeastern Equatorial Pacific from the SIO Sediment Data Bank, November 1978, indicated by the heavy black line, and as reported by Bastien-Thiry et al. (1977), indicated by the shaded area.

This sounds quite low to persons who have been thinking in terms of trillions of tons, but even this amount of nodules can provide a very significant source of metals. Table 3 shows the amounts of metals that could be recovered from nodules in the Clarion-Clipperton Zone, given different levels of mining efficiency. Nickel from nodules in this region amounts to 12-87% of world land reserves, and cobalt is 50-400% of land reserves.

These, of course, are only the demonstrated paramarginal resources. We have enough data to assure ourselves that they exist, and their grade and abundance make them likely to be economic sources of raw materials within the next decade. The nodule resources so far unexplored, including those that might not be economically mined until much later, are much larger.

ESTIMATING UNEXPLORED NODULE RESOURCES:
GEOLOGIC FACTORS

For estimating the magnitude of unexplored nodule resources we must rely to a certain extent on geologic information that indicates whether or not an area is likely to be favorable for the formation of sufficiently concentrated nodule deposits containing the required percentages of nickel, copper, cobalt, and manganese. Therefore, before we consider where these resources might be and what amount of metals they might contain, let us review current knowledge about relevant geological factors.

It has long been established in the literature that nodules rich in nickel and copper form only at great depths in the ocean. Virtually all nodules containing more than 2.25% nickel plus copper occur at depths between 4000 and 6000 meters. Submarginal nodule resources with grades between 1.5 and 2.25% nickel plus copper can also occur in water as shallow as 3600 meters.

Cobalt content, on the other hand, is well known to show an inverse correlation to water depth.

Table 3
RECOVERABLE METALS IN PARAMARGINAL DEMONSTRATED
RESOURCES (Clarion-Clipperton Zone)

	In situ amount	Amount recoverable at mining efficiency of: 20%	40%	Estimated land reserves[a]
Nodules[b]	4000-15,000	560-2100	1120-4200	
Nickel[c]	35-131	6.3-24	13-47	54
Copper[d]	29-108	5.2-19	10-39	460
Cobalt[e]	6.4-24	0.8-2.9	1.5-5.8	1.5
Manganese[f]	706-2600	120-450	240-900	2000

All values in millions of tonnes.

a. Archer (1978).
b. In situ nodules contain 30% water. Amount of recoverable nodules calculated on a dry-weight basis.
c. Nodules estimated to contain 1.25% nickel; processing recovery efficiency 90%.
d. Nodules estimated to contain 1.03% copper; processing recovery efficiency 90%.
e. Nodules estimated to contain 0.23% cobalt; processing recovery efficiency 60%.
f. Nodules estimated to contain 25.2% manganese; processing recovery efficiency 85%.

Nodules with cobalt content above 0.4% are generally associated with volcanic seamounts, with the nodules containing the most cobalt found on the tops of the seamounts (Arrhenius et al., 1978).

Mero (1965) was the first to point out that deposits of abundant nodules are most frequently found in areas where sedimentation rates are low. These are generally areas where the sediment is pelagic clay or siliceous ooze, but high concentrations

can also occur where sedimentation is inhibited as a
result of current action (Cronan, 1977). (Siliceous
oozes are sediments containing a large component of
the remains of siliceous organisms. Radiolaria are
the predominant siliceous organism near the Equator;
elsewhere, diatoms predominate.) Nodules are seldom
found in association with terrigenous sediments,
probably because incipient nodules are buried by
large influxes of material from the continents (Horn
et al., 1972), or perhaps because suitable nuclei are
not available (Cronan, 1977). Except on seamounts,
nodules are usually absent or sparse where the sediment is calcareous ooze. This effect seems to be due
to the reduction of manganese to divalent state at
the high rate of deposition and decomposition of
organic matter in these sediments (Arrhenius et al.,
1978).

In some areas the seafloor is covered with a thick
manganese pavement or crust rather than individual
nodules. Although such pavements form quite abundant
deposits, they are generally poor in nickel and
copper and should not be considered part of the
minable resource.

Horn et al. first suggested that the equatorial
radiolarian ooze sediments are particularly favorable
for the growth of nodules with high nickel and copper
contents. Table 4 confirms this observation. The
reason for high nickel in this area has not been com-
pletely determined; perhaps the porosity of these
sediments is an important factor as they suggested.
We are convinced, however, that the high copper content in nodules from such areas is due to the concentration of this metal in the skeletons of radiolaria
(Arrhenius, 1963).

Many authors have reported a strong correlation
between nickel and copper in nodules, and this led us
to believe that the average Cu/Ni ratio of 0.84 found
in the Clarion-Clipperton Zone was likely to be typical of the entire ocean. On the contrary, average
Cu/Ni is only 0.58 for the rest of the ocean. Nodules

Table 4
Association between sediment type and nodule grade (for 478 locations where information on both is available)

Sediment	Number of Stations	Percentage of stations within each grade range			
		Ni+Cu 1%	Ni+Cu 1-2%	Ni+Cu 2-2.5%	Ni+Cu 2.5%
Pelagic clay	206	58	24	11	7
Equatorial siliceous ooze and clay	98	12	21	31	36
Other siliceous ooze and clay	9	78	22	0	0
Calcareous ooze	97	86	12	2	0
Terrigenous material	68	85	15	0	0

high in nickel occur in many different regions
(Figure 6); however, as expected theoretically,
nodules greatly enriched in copper occur only near
the Equator (Figure 7).

Nickel values higher than 2% have been reported
in the Clarion-Clipperton Zone, but most nodules in
that region contain nickel in the range 1.2-1.6%,
with a mean of 1.25%. Unless nickel values are
unusually high or cobalt content is high enough to
make up for the lack of copper, nodule deposits outside the equatorial regions are unlikely to reach
paramarginal resource grade (expressed as a nickel-
equivalent grade, Ni + Cu/3 + 2Co).

Manganese is another important raw material that
can be obtained from seafloor nodules. The total
amount of manganese that is contained in nodules is
enormous, probably amounting to more than present
land reserves. Some nodules and crusts with especially
high manganese content (more than 35%) have been found,
but these contain almost no nickel or copper. The
abundance and extent of such deposits have not been
studied. Otherwise, however, the nodules that are
high in manganese (24-30%) are the same nodules that
contain high nickel and copper. Most nodules outside
of the eastern Equatorial Pacific contain manganese
in the 10-20% range. Although a time may come when
such nodules can be mined economically for their
manganese, that time is surely very far in the future.

Another factor that needs to be taken into account
in estimating worldwide nodule resources is the inverse correlation between nickel plus copper grade
and nodule abundance (Menard and Frazer, 1978).
Based on data now available for more than 250 locations at which both grade and concentration have been
determined, the correlation coefficient between these
two variables is -0.43, at the 99.999% confidence
level (figure 8). This means there is one chance in
100,000 that grade and abundance are really uncorrelated and that the apparent inverse relationship
was produced by sampling error. For the Clarion-

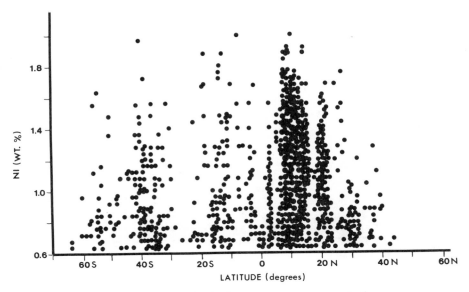

Figure 6.   Percent nickel in manganese nodules as
a function of latitude.

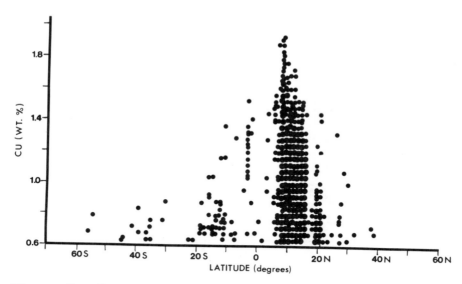

Figure 7.   Percent copper in manganese nodules as a function of latitude.

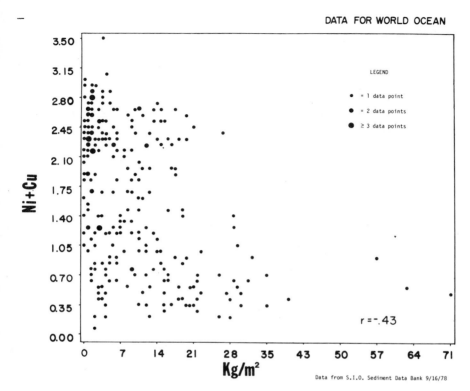

Figure 8.   Percent nickel plus copper in manganese
nodules as a function of seafloor concentration.

Clipperton area the correlation coefficient is less,
-0.22 at the 98% confidence level.

One area for which our data show no correlation
between grade and abundance is DOMES Site C, within
the Deepsea Ventures claim site. Sorem et al. (1978)
have just completed a study of box cores collected by
the R/V Oceanographer from Site C, however, and they
found that the box cores which collected the greatest
mass of nodules show relatively low nickel plus
copper grade. They report that total weights are
inversely correlated with weighted average nickel
plus copper value of each box core at the level of
-0.48.

What is the practical effect of such an inverse
relationship on resource prediction? I divided our
data into seven categories according to nickel plus
copper content and calculated the mean concentration,
in kg/m^2, for each grade interval. Results are shown
in figure 9. Concentrations vary greatly within each
grade interval, with standard deviations about the
same as the mean value for each interval. Nevertheless, an analysis o- variance shows that the varia-
tion among the means is significant at much better
than the 99% confidence level.

Suppose one chooses to ignore this relationship
between grade and abundance in estimating nodule
resources. How much error is likely to be introduced? If we assume that grade and abundance are
independent, we can pick any level for each parameter
and make a calculation like the following: If 10%
of the locations have the required grade and 10%
have the required abundance, then we can multiply
these numbers together and assume that 1% of the
locations (and thus presumably 1% of the area
involved) would meet both criteria. This is the
global estimator method used by some authors.

Let us try this out on the actual data and some
subsets of it. Thirty-six percent of the stations
for which both grade and concentration are known
have nickel plus copper greater than 2.25%, and 56%

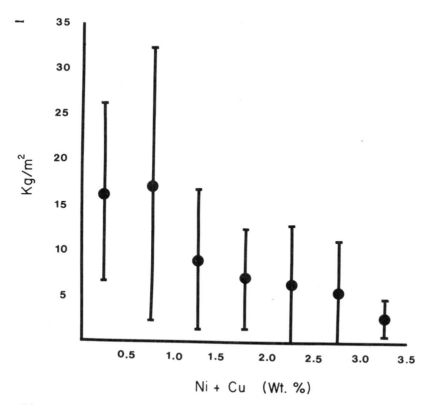

Figure 9. Average seafloor nodule concentration for different nickel plus copper grade intervals.

of the stations have concentrations greater than 5
kg/m^2. Assuming independence of grade and abundance
we would estimate that 20% meet both criteria, but
actually only 14% do. We have overestimated by 42%.
If we make the same calculation with a grade of 1.8%
or concentration of 10 kg/m^2, the results are
similar--a 50% overestimate of the area suitable for
mining.

For an area southwest of Hawaii studied by the
Japanese Geological Survey (Mizuno and Moritani,
1977), an assumption that grade and abundance are
uncorrelated would lead to overestimating the fraction of the area that would meet both criteria by
factors ranging from 2.5 to 5, depending on the grade
and concentration limits chosen. In the Clarion-
Clipperton Zone as a whole, however, the over-
estimation would be only 1-15%, again depending on
the limits.

In summary, in predicting nodule resources we can
supplement our actual data with the following geologic
knowledge:

1. Paramarginal resources can be expected near
the Equator in areas where the water is 4000-6000
meters deep, sedimentation rates are slow, and the
sediment is radiolarian ooze or radiolarian clay.
All such areas can probably be considered at least
submarginal resources. Nodule concentration rather
than grade is likely to be the factor limiting paramarginal resources to only portions of such areas, at
least in the Pacific Ocean.

2. Other areas with water 3600-6000 meters deep
and slow sedimentation rates are likely to contain
moderate or highly concentrated nodule deposits. The
necessary nickel plus copper or nickel plus cobalt
grades, however, are probably limited to the areas
where pelagic clay is the underlying sediment. Such
deposits can contain nodules with enough nickel and
copper (or cobalt) to qualify as paramarginal
resources, but they are more likely to be part of the
submarginal resources.

3. Shallow water areas where the sediment is
calcareous ooze could include resource deposits if
cobalt were the main target for mining. Otherwise,
the chance that areas of calcareous ooze or terrigenous sediment will contain minable nodule deposits
is practically nil.
4. Since nodule grade and abundance are inversely
correlated, we cannot use any method of resource
estimation that relies on their being independent
variables. We must also expect the most highly concentrated nodule deposits to be relatively low in
copper and nickel. /

UNEXPLORED PARAMARGINAL RESOURCES
When we move outside of the northeastern Equatorial
Pacific, we are in regions where the grade data are
much sparser and measured nodule concentrations are
very few indeed. The SIO Sediment Data Bank contains
nodule assays for only 1300 locations outside the
Clarion-Clipperton Zone, and measured nodule concentrations for only 150 sites. The unexplored regions
offer ample scope for authors to use their imagina-
tions, and reported resource estimates may reveal
more about the innate optimism or pessimism of the
individual authors than about the magnitude of the
resource.
     Four authors have attempted to estimate para-
marginal resources systematically for the entire
ocean. Both Archer (1976) and Holser (1976) used
two different methods--the global estimator and the
prime area approach. The global estimator assumes
that grade and abundance are uncorrelated and that
data were collected randomly. The method involves
estimating the fraction of the total seafloor that
has sufficient nodule abundance and the fraction that
contains nodules of the required grade. The product
of these two fractions is assumed to represent the
portion of the seafloor that meets both requirements.
     Using this method Archer estimated 44 billion
tonnes and Holser estimated 85 billion tonnes of

nodules outside the Clarion-Clipperton Zone in
deposits that would meet the grade and abundance
criteria for paramarginal resources. Since we have
shown that the independence of grade and abundance
is an untenable assumption and that these two parameters are inversely correlated, we are certain that
these are overestimates. Note that Menard and Frazer
(1978) did not suggest any numerical relationship
between grade and abundance that could be used in
global-estimator-type calculations. Such a relationship would have to include other variables, such as
sediment type. On the basis of present knowledge it
is impossible to determine any simple formula that
can appropriately be applied to worldwide or ocean-
wide frequency distributions to predict nodule
resources.

We must, therefore, examine the data for
individual regions and take their geological characteristics into account, as Archer and Holser did in
their prime area estimates. Their paramarginal
resource predictions by this second method are, as
expected, considerably less than with the global
estimator: 11.4 billion tonnes for Archer and 28
billion tonnes for Holser.

With my grid estimator method I only located one
potentially paramarginal area outside the Clarion-
Clipperton Zone, an area in the Peru Basin containing
an estimated 3.5 billion tonnes of nodules (Frazer,
1977). My estimate is probably too low, because
averaging grade over five-degree squares eliminated
some smaller areas with high average grade that might
contain enough nodules for a first-generation mine
site.

Very tentative results reported by Pasho (1977)
indicate a probable range of 5 to 47 billion tonnes
outside of the northeastern Equatorial Pacific that
could be mined before the end of this century. The
probability that this resource would be as great as
47 billion tonnes is only 5%.

In order to evaluate these estimates, let us
examine the data for some specific regions that may
contain paramarginal nodule resources. The most
promising is the southeastern Equatorial Pacific,
particularly the Peru Basin mentioned above. Because
of the East Pacific Rise and some other features,
this region is not quite the mirror image of the
Equatorial North Pacific as some have supposed. It
does, however, contain some areas of radiolarian ooze.

The French AFERNOD group has done considerable
prospecting in the South Pacific. Perret (1972)
expressed their hopes of finding workable deposits
within the territorial waters around the Polynesian
archipelago. Although the French have not yet
reported on their South Pacific results, they
apparently found no mine sites there. At the UN
Workshop on Seabed Mineral Assessment, Bastien-Thiry
stated that they had found high-grade nodules in the
South Pacific as well as areas of abundant nodules,
but they did not find areas where the two coincided.

The radiolarian ooze areas south of the Equator
in the Indian Ocean are generally considered to be
likely areas for paramarginal resources (Archer,
1976; Holser, 1976), and according to our geologic
theories they should be. We recently studied these
areas as part of a preliminary report on nodule
resources in the Indian Ocean (Frazer et al., 1978).
A very few samples showed copper plus nickel grades
above 2.25%, but samples from nearby sites were of
lower grade. Since we believe (see table 1) that
a few assays within an area can give a reasonable
indication of the true average grade for the area,
we were forced to conclude that nodule resources in
this region and in the rest of the Indian Ocean are
probably submarginal. If we had developed a statistical method that allowed us to estimate probabilities,
we would likely have found some probability less than
50% that the Indian Ocean contains paramarginal
deposits.

A similar situation exists in the Cape Basin of
the South Atlantic, where Holser identified 6.2
million km^2 of prime area, of which some portion
would meet the criteria for first-generation mine
sites. First of all, if we include only the areas of
pelagic clay sediment and not the surrounding areas
of calcareous ooze where we have shown that nodules
are unlikely to be either high-grade or abundant,
the Cape Basin covers little more than two million
km^2. Twenty nodule assays are available for this
region, averaging 1% nickel plus copper. Two assays
are barely above the 2.25% grade required for paramarginal resources; three are within the submarginal
grade range, and the rest are lower. Again, it seems
reasonable to conclude that there is a small but
finite probability that the Cape Basin contains at
least enough paramarginal nodule deposits for one
mine site, and a good probability that there are submarginal resources here. Only Pasho used a method
that calculated such probabilities and included them
in the estimated resource range.
    We conclude that Pasho's estimate of 5-47 billion
tonnes nodules, which includes Archer's and Holser's
prime area estimates, is the most satisfactory prediction of unexplored paramarginal resources. The
actual paramarginal resources outside of the Clarion-
Clipperton Zone are most likely to be at the lower
end of this range, which would mean that they are
approximately equal to the demonstrated paramarginal
resources of 4-15 billion tonnes. Thus we can double
the figures in table 3 to obtain an estimate of total
paramarginal resources and the metals that might be
recovered from them.
    Some profitable by-product of nodule mining might
bring these deposits into the economic category. As
we have seen, the nodule deposits richest in copper
and nickel form on a substrate of siliceous radiolarian sediments. The skeletons of the radiolaria,
which are mixed with pelagic clay and comprise 10-50%
of the sediment, are themselves a valuable raw

material. These microscopic organisms have skeletons
composed of opaline silica in a fibrous molecular
arrangement which results in high elasticity and
strength (figure 10). The strength, porosity, per-
meability, and other physical properties of the
radiolarian silica is for many applications superior
to those of diatoms, another type of siliceous
microorganism that is widely used in industrial
applications.

Arrhenius has developed processes for transforming
radiolarian ooze into insulating, highly porous,
lightweight ceramics with strength that at comparable
densities is considerably higher than any economically
and structurally comparable product on the market.
Other potential applications for radiolarite include,
acoustic and thermal insulation in airplanes, where
low weight and absence of fume hazard are necessi-
ties; catalyst carriers; fillers and extenders in
paints and plastics; and fillers in acetylene gas
tanks. Arrhenius is presently completing an economic
evaluation of deepsea radiolarite sediments under a
grant from the U.S. Bureau of Mines.

Although we cannot say at this time how simultaneous recovery of radiolarite would affect the
economics of seafloor nodule mining, it is something
to keep in mind. It could turn a problem--the
unavoidable transport of some sediment to the surface
along with the nodules--into an asset.

SUBMARGINAL RESOURCES
Submarginal nodule resources are defined as those
deposits which do not meet the criteria for first-
generation mining but which may become economically
recoverable at some time in the future. I have
suggested that these would include deposits where
average nodule grade is 1.5-2.25% nickel plus copper
as well as deposits of higher grade with seafloor
concentrations less than 5 kg/m^2.

Archer is the only author who has attempted to
predict the amount of submarginal nodule resources.

Figure 10. Radiolarian skeletons.

Using the global estimator method in 1976 he predicted
143-218 billion tonnes of nodules in situ with nickel
plus copper grade higher than 0.88% (one-half the
cutoff grade for paramarginal resources) and concentrations greater than or equal to 5 kg/m^2. More
recently (1978) he predicted 235 billion tonnes as a
"theoretical (and very approximate) maximum" for
deposits with grades higher than 0.88% and minimum
concentrations of 2.5 kg/m^2.

From a brief review of my original grid estimator
calculations I find that nodule deposits with average
grade ranging from 1.5 to 2.25% nickel plus copper
amount to something on the order of 100-200 billion
tonnes. This should not be taken as a definitive
estimate. We are now attempting to make a more
reliable prediction of submarginal resources in addition to unexplored paramarginal resources based on
the geological characteristics of various areas as
well as on the existing data. There seem to be
enough nodule assays and a sufficient understanding
of the relationship between nodule composition and
geological factors for us to make reasonable
estimates of nodule grade for most areas, although
we would expect our estimates to be improved as
further information appears. However, there is a
serious lack of nodule concentration measurements.

Before we spend too much time calculating the
size of submarginal resources, we must remind ourselves that the minimum average grade of 1.5% nickel
plus copper is merely the suggestion of a few people
who got together briefly and tried to make a
reasonable guess. No economic studies have been done
to indicate when, if ever, such deposits might
become profitable to mine.

The simple requirements we have specified for
submarginal resources fail to take into account the
expected Cu/Ni or Co/Ni ratios of such deposits or
possible changes in the relative values of these
metals. The possibility that shortages of cobalt or
manganese at some time in the future might make one

of these metals the prime target for nodule miners,
with nickel and copper of only secondary interest,
has not been considered in published resource estimates. Neither have we taken into account the value
of other metals, such as molybdenum or vanadium,
which occur as trace elements in nodules.

Furthermore, grade and abundance are not the only
factors that will determine whether a nodule deposit
is economically minable. Although I expect that the
mining companies have made careful studies of all the
factors involved, I have not seen any published work
on possible trade-offs between various factors. For
example, does especially high grade compensate for
low concentrations? Would a lower grade or abundance
be acceptable if costs were reduced by mining in
shallower water or closer to land? Can high cobalt
content make up for low copper in nodules far from
the Equator, or would mining such nodules merely
produce a surplus of cobalt? These are questions for
economists and mining engineers, not for geologists.
The answers to such questions are important for predictions of paramarginal as well as submarginal
resources.

Even if submarginal nodule deposits amount to
only 50 billion in situ tonnes containing an average
of 1% nickel and 15% manganese, they would contain
several times the estimated submarginal land
resources of these metals. Cobalt content would be
many times land resources, but copper would probably
be negligible. Only a portion of these nodules could
be recovered, however, even if the deposits do become
economically minable.

CONCLUSIONS
Demonstrated paramarginal resources in the region of
the northeastern Equatorial Pacific between the
Clarion and Clipperton Fracture Zones are estimated
to contain 4-15 billion tonnes of nodules. This
range cannot be narrowed further without additional
data. These are resources for which the quality and

quantity have been computed mostly from sample
analyses and measurements. They are expected to meet
the grade and abundance requirements for first-
generation mining.

How many tonnes of metals can actually be
recovered from such deposits depends on the mining
efficiency, for which estimates vary between 10 and
50%. At 24% mining efficiency (Pasho's estimate of
the value at the 50% probability level) and production requirements of 75 million dry tonnes per mine
site, these deposits would provide 9-34 mine sites
from which 8-28 million tonnes nickel, 6-23 million
tonnes copper, 0.9-3.5 million tonnes cobalt, and
144-530 million tonnes manganese could be recovered.

Unexplored paramarginal resources outside the
Clarion-Clipperton Zone are estimated to contain
5-47 billion tonnes of nodules in situ; I believe
that the lower end of this range is most probable,
that is, that they contain about the same amounts of
metals as the demonstrated resources. These resources
include deposits which may meet grade and abundance
requirements for mining within the next decade but
for which relatively few measurements are available.

Paramarginal seafloor nodule resources, demonstrated and unexplored together, are thus likely to
provide a significant source of metals. Except for
cobalt, however, they are not likely to be a resource
enormously greater than land reserves. Nickel from
seafloor nodule deposits suitable for first-generation
mining may amount to as much as 1.5 times land
reserves, but there is some probability that it could
be only a fourth of land reserves. Even if the most
optimistic predictions of mining efficiency prove to
be correct, the paramarginal nodule resource is quite
unlikely to be large enough to provide hundreds of
mine sites.

Submarginal nodule resources, which might be
economically mined within the next century as the
metals they contain become scarcer, are almost certainly larger than the paramarginal resources and as

S—

large as land resources. Work is presently being done to estimate their size. Such estimates would be more useful, however, if there were wider agreement on the required grade, abundance, and other factors necessary to define submarginal resources.

ACKNOWLEDGMENTS
The SIO Sediment Data Bank and my work on evaluation of manganese nodule resources is supported mainly by the U.S. Bureau of Mines under Grant No. GO-264024. Mary B. Fisk provided extensive assistance in the preparation of this paper.

REFERENCES

Archer, A. A., 1976. Prospects for the exploitation of manganese nodules: The main technical, economic and legal problems. In: Papers Presented at the IDOE Workshop, Suva, Fiji, 1-6 September 1975, G. P. Glasby and H. R. Katz, ed. (Tech. Bull. 2. UN Economic and Social Commission for Asia and the Pacific, CCOP/COPAC, 21).

Archer, A. A., 1978. The possibilities for manganese nodule mining. Presented at the International Ocean Development Conference and Exhibition, September 1978, Tokyo, Japan.

Arrhenius, G., 1963. Pelagic sediments. In: The Sea, 3, M. N. Hill, ed. New York: Interscience Publishers.

Arrhenius, G., 1977. New ceramic materials from deep-sea resources. Scripps Institution of Oceanography Ref. No. 77-27, June 1977.

Arrhenius, G., K. Cheung, S. Crane, M. Fisk, J. Frazer, J. Korkisch, T. Mellin, S. Nakao, A. Tsai, and G. Wolf, 1978. Counterions in marine manganates. Scripps Institution of Oceanography Ref. No. 78-26.

Bastien-Thiry, H., J. P. Lenoble, and P. Rogel, 1977. French exploration seeks to define mineable nodule tonnages on Pacific floor. Engineering/Mining Journal (July).

Cronan, D. S., 1977. Deep-sea nodules: Distribution and geochemistry. In: Marine Manganese Deposits, G. P. Glasby, ed. Amsterdam: Elsevier.

Ewing, M., D. Horn, L. Sullivan, T. Aitken, and E. Thorndike, 1971. Photographing manganese nodules on the ocean floor. Oceanology International, 6: 12.

Frazer, J. Z., 1979. The reliability of available data on element concentrations in seafloor manganese nodules. In: Manganese Nodules: Dimensions and Perspectives. Dordrecht: Reidel.

Frazer, J. Z., and G. Arrhenius, 1972. World-wide distribution of ferromanganese nodules and element concentration in selected Pacific nodules. Tech. Rep. Off. Int. Decade Ocean Explor. 2 (51 pp.).

Frazer, J. Z., M. B. Fisk, J. Elliott, M. White, and L. Wilson, 1978. Availability of copper, nickel, cobalt and manganese from ocean ferromanganese nodules. Scripps Institution of Oceanography, Ref. No. 78-28, September 1978.

Holser, A. F., 1976. Manganese nodule resources and mine site availability. Professional Staff Study, Ocean Mining Administration, U.S. Dept. of Interior, Washington, D.C. Unpublished, 12 pp.

Horn. D. R., B. M. Horn, and M. N. Delach, 1972. Ferromanganese deposits of the North Pacific. Tech. Rep. Off. Int. Decade Ocean Explor. 1 (78 pp).

Kaufman, R., 1974. The selection and sizing of tracts comprising a manganese nodule ore body. OTC 2059, Offshore Tech. Conf. Preprints, 11: 283.

Kildow, J. T., M. B. Bever, V. K. Dar, and A. E. Capstaff, 1976. Assessment of economic and regulatory conditions affecting ocean minerals resource development. Report to the U.S. Dept. of Interior. Unpublished.

McKelvey, V. E., N. A. Wright, and R. W. Rowland, 1978. Manganese nodule resources in the northeastern Equatorial Pacific. (In press.)

Menard H., and J. Z. Frazer, 1978. Manganese nodules on the sea floor: Inverse correlation between grade and abundance. Science 199: 969.

Mero, J., 1965. The Mineral Resources of the Sea. Amsterdam: Elsevier.

Mero, J., 1977. Economic aspects of nodule mining. In: Marine Manganese Deposits, G. P. Glasby, ed. Amsterdam: Elsevier.

Mizuno, A., and T. Moritani, eds., 1977. Deep-sea mineral resources investigation in the central-eastern part of central Pacific basin, January-March 1976 (GH76-1). Geol. Sur. Jap., Cruise Rep. No. 8.

Pasho, D. W., and J. A. McIntosh, 1976. Recoverable nickel and copper from manganese nodules in the northeast Equatorial Pacific--preliminary results. Canadian Institute of Mining and Metallurgy Bulletin, 69 (773): 15.

Pasho, D. W., 1977. Review of the development of deep seabed manganese nodules. Northern Miner, April 14, B6, B9, B13.

Pasho, D. W., 1979. Determining deep seabed mine-site area requirements: A discussion. In: Manganese Nodules: Dimensions and Perspectives. Dordrecht: Reidel.

Perret, J., 1972. Polymetallic nodules on the ocean floor. In: Report of the Ninth Session, the Committee for Co-ordination of Joint Prospecting for Mineral Resources in Asian Offshore Areas (CCOP) and Report of the Eighth Session of its Technical Advisory Group together with Part 2: Technical Documents, held in Bandung, Indonesia, 1972.

Schultze-Westrum, H.-H., 1973. The station and cruise pattern of the R/V Valdivia in relation to the variability of manganese nodule occurrences. In: Papers on the origin and distribution of manganese nodules in the Pacific and prospects for exploration, M. Morgenstein, ed. Honolulu, July, 145.

Sorem, R. K., W. R. Reinhart, R. H. Fewkes, and W. D. McFarland, 1978. Occurrence and character of manganese nodules in the "Horn Region" east equatorial Pacific Ocean. In: Pacific Ocean Nodule Symposium, 3, J. L. Bischoff, ed. (In press.)

Spanger, M. B., 1970. Deep sea nodules as a source of copper, nickel, cobalt, and manganese. In: New Technology and Marine Resource Development. New York: Praeger.

THE ROLE OF DEEP SEABED MINING IN THE FUTURE SUPPLY OF METALS

Lance Antrim

Deep ocean mining has been proposed as a supplement
or alternative to land-based sources of nickel,
copper, cobalt, and manganese, all basic industrial
raw materials important to the continued growth of
the economy as well as to the nation's security.
The importance of ocean mining to the nation can be
examined in terms of the land-based sources that
might be replaced by deep ocean mining and the effect
of such replacement on the availability of minerals
to U.S. industries. The potential role of ocean
mining is discussed in four sections. First, the
projected demand for the four metals through the end
of the century is summarized. Second, the future
land-based sources are estimated from the distribution of current mine production and the distribution
of ore reserves. Third, the resource potential of
the known nodule minesites is examined in light of
the future demand for the metals and the alternative
land-based sources. Finally, conclusions are drawn
as to the national interest in ocean mining
development.

FUTURE DEMAND FOR NODULE METALS

Prediction of the future metal requirements of the
United States and the world is difficult and prone to
error. Extrapolation of historical growth cannot
account for changes in consumer preference or new
technological developments, nor can it predict the
effect of increasing industrialization of developing
states. In light of these uncertainties, the U.S.
Bureau of Mines has prepared a range of estimates of
the rate of growth of consumption of the four nodule
metals through the rest of the century. These
estimates are shown in table 1.

The estimated rates of growth of consumption of
the four metals can be used to estimate both the
annual and the cumulative consumption of the metals.

Table 1
Estimated average annual rates of growth of
metal consumption, 1975 to 2000

		Low	Probable	High
Nickel	U.S.	1.7%	3.0%	3.3%
	World	2.2%	3.4%	3.8%
Copper	U.S.	1.9%	3.0%	4.1%
	World	2.6%	3.7%	4.7%
Cobalt	U.S.	1.3%	2.9%	3.7%
	World	1.8%	3.2%	4.0%
Manganese	U.S.	1.2%	1.7%	2.6%
	World	2.5%	2.9%	3.8%

Source: U.S. Bureau of Mines.

These estimates are shown in table 2. The estimates
assume the rates of growth used in table 1; they are
extrapolated from values of consumption for a base
year (1975, except in the case of cobalt which is
based on 1976). The base value is calculated from a
trend line based on the metal consumption during the
preceding 20 years.

LAND-BASED RESOURCES
The future supply of nickel, copper, cobalt, and
manganese can be estimated from three types of
information: development plans for the construction
of new facilities, estimates of ore reserves, and
estimates of known and speculative ore resources
throughout the world. The construction of new mining
and processing facilities takes about five years, so
for that period new mineral capacity may be predicted
from public information about the progress of the
facilities. Projects that are in the conceptual
stage and the future expansion of existing facilities
may not be reported in the current development plans

Table 2
Forecast of U.S. and world consumption of nodule metals through the
year 2000 (thousands of short tons)

		Base Year	Consumption in base year	2000 forecast range Low	Probable	High
Nickel						
U.S.:	Primary	1975	154	300	415	450
	Cumulative			6,200	7,400	7,800
World:	Primary	1975	670	1,290	1,705	1,900
	Cumulative			26,100	29,400	31,800
Copper						
U.S.:	Primary	1975	1,307	2,700	3,500	4,600
	Cumulative			54,000	63,000	74,000
World:	Primary	1975	7,114	14,700	19,500	24,700
	Cumulative			274,000	323,000	374,000
Cobalt						
U.S.:	Primary	1976	9.7	14	20	24
	Cumulative			282	349	389
World:	Primary	1976	37.2	58	80	96
	Cumulative			1,138	1,369	1,524
Manganese						
U.S.:	Primary	1975	1,133	1,910	2,130	2,680
	Cumulative			41,000	44,000	50,000
World:	Primary	1975	10,810	20,000	22,100	27,600
	Cumulative			376,000	398,000	459,000

Forecasts for the year 2000 are based on growth rates reported in table
1 and on values calculated for the base year based on trends for the
preceding 20 years. The trend values are:

	U.S.	World
Nickel	198	744
Copper	1,682	7,811
Cobalt	9.9	37.5
Manganese	1,394	10,831

of the mining companies. These project
exploit ore deposits that are known to
so they will be located in areas that
reserves of ore. Finding new deposit
range process, often taking ten years or mo.
it is likely that new mineral development will be i..
areas of known reserves for the next decade. Since
market conditions and mining laws are subject to
change, the estimates of reserves must be supplemented
by resources that are marginally uneconomic or that
are currently unminable for legal or political
reasons. Finally, for the more distant future, the
estimates of reserves must be supplemented by the
remainder of the metal resources, including hypothetical and speculative resources, to indicate
possible areas of mineral development, even though
specific information about the size and cost of the
development cannot be determined. The outlook for
the future production of nickel, copper, cobalt, and
manganese is described below.

Nickel

Nickel currently is mined in a dozen nations outside
of the communist states. Since the opening of the
Sudbury deposit of nickel sulfide ore in Ontario,
Canada, fifty years ago, Canada has been the largest
producer of nickel in the world. As nickel consumption has increased during the past twenty years, the
Canadian production has been supplemented by production in other nations. The most notable contribution
has been from New Caledonia, a French territory in
the Southwest Pacific. Unlike the Sudbury deposit,
the nickel ore of New Caledonia is a nickel laterite.
These deposits are mined using surface mining tech-
niques and are processed into an intermediate product
that is shipped elsewhere for further processing.
Generally, nickel laterite deposits are less profitable than the Canadian sulfide deposits, but the
large reserves of high-grade ore and the relative
simplicity of the mining methods have made New

aledonia the second largest producer of nickel.
Other nations in the southwest Pacific also have
large production of nickel. Australia, with both
sulfide and laterite deposits, is a major producer.
The combined production of Australia, New Caledonia,
and Canada, which are all dependable suppliers to
the developed nations, accounted for 73 percent of
the 1975 production of nickel from noncommunist
states.[1] This fraction will decline in the future,
however, because the trend in nickel production is
toward the construction of new capacity in developing
states. Indonesia, the Philippines, Guatemala,
Greece, the Dominican Republic, Botswana, and South
and Central America, which together produced 19% of
the noncommunist output of nickel in 1975, account
for 34% of the projected 1980 noncommunist nickel
capacity.[2]

Development of nickel in the next decade will
follow the patterns set by the exploration programs
of the mining companies in the recent past. Production from the large reserves of Canada and New
Caledonia will continue to show some increase, but
the trend of the exploration programs has been to the
nations of the western Pacific and Central and South
America. The fruits of these exploration programs
are illustrated in table 3, which shows the distribution of world reserves of nickel. Canada, New
Caledonia, Cuba, and the Soviet Union have large
reserves, but these have been known for some time.
The reserves of Australia, Indonesia, the Philippines,
and the countries of Central and South America
(Guatemala, Dominican Republic, Brazil, Colombia),
however, are the result of more recent exploration
and represent the areas that the mining companies
will develop in the future.

Copper
Of the many countries in which copper is mined, the
largest producer, accounting for 18% of the 1975
world mine production, is the United States.[3] The

Table 3
World nickel reserves

	Reserves (thousand short tons)	Percentage of total world reserves*
North America		
Canada	9,600	16%
United States	200	1%
Total	9,800	16%
Central and South America		
Dominican Republic	1,100	2%
Colombia	900	2%
Guatemala	300	1%
Brazil	200	1%
Total	2,500	4%
Oceania		
Australia	5,600	9%
Indonesia	7,800	13%
New Caledonia	15,000	25%
Philippines	5,400	9%
Total	33,800	56%
Africa (Total)	2,300	4%
Communist States		
USSR	8,100	14%
Cuba	3,400	6%
Total	11,500	19%
World Total	59,900	100%

*Fractions are rounded to nearest integer.

Source: U.S. Bureau of Mines.

Soviet Union is also a large producer, but this metal
reaches the copper market only in small amounts.
Several developing states also make a large contribution to the world supply of copper. In 1975, Chile
supplied 14% of the noncommunist copper supply of
6,533 tons, while Zambia supplied 11% and Zaire
supplied 8%.[4] These nations, along with Peru,
Indonesia, and Mauritania, form a copper producers'
association known as CIPEC which claims to control
about 38% of the world mine production and 72% of the[5]
export trade of mine and smelter products. The CIPEC
nations have a large influence on export trade
because the production of the United States is mainly
consumed domestically, leaving the rest of the
developed world heavily reliant on developing states
for copper. The power of CIPEC to influence prices
may be less than suggested by these figures, however,
because of the needs of the individual states to
maintain current income to support their debts and to
maintain their development projects. For these
reasons, the CIPEC nations are reluctant to reduce
production in order to obtain higher prices in the
near future.
    Examination of the distribution of production
capacity over time indicates that the trend to
increased copper production in developing states is
likely to accelerate because of the rising costs of
domestic production for U.S. mines. These costs
result from a variety of factors, including the lower
average ore grade and higher labor costs of U.S.
mines, but a large chunk is due to U.S. antipollution
standards. Air quality standards are estimated by
the U.S. Bureau of Mines to have added over 10% to
the cost of domestic copper in 1970.[6] A 1978 study
by A. D. Little, Inc., comparing the relative costs
of meeting the National Ambient Air Quality Standards
(NAAQS) and a proposed stiffer set of regulations
projected that the stiffer regulations would add
29-39% to costs through 1987.[7] The report estimated
that the additional regulation would reduce domestic

production in 1987 by 25-33% compared to production under NAAQS.[8] The higher costs of domestic production have resulted in increased interest in foreign investment by the copper industry, and this trend is likely to continue through the next decade.

Worldwide copper reserves are estimated at 503 million tons.[9] The United States accounts for 93 million tons, about 18% located primarily in the states of Arizona, Utah, New Mexico, Montana, and Michigan.[10] Chile is also credited with reserves of 93 million tons.[11] Available reserves are shown in table 4. This table shows that mining operations could be started or expanded in virtually any region of the world, with the exception of Europe. Even Oceania, including Papua New Guinea, Australia, and the Philippines, has reserves equal to two and a half times the 1975 total world production of copper. In the future, some development will occur in the United States, but domestic pollution control measures will probably make foreign operations more attractive for investment than many of the domestic opportunities. It is likely that the expansion of the copper industry over the next ten years will be spread over a range of developing countries including Chile, Peru, Zambia, and Zaire.

Cobalt
With the exception of a single mine in Morocco, cobalt is mined only as a by-product of nickel or copper. The largest production of cobalt is from the African copper belt that runs through Zaire, in the province of Shaba, and Zambia. The deposits of cobalt-rich ore have made Zaire the world's dominant producer of cobalt; Zambia is in second place with a considerably lower production. Zaire alone accounted for 59% of the 1975 noncommunist production of cobalt, while the African nations of Zaire, Zambia, and Morocco accounted for 76%.[12] The remainder of the world cobalt supply is produced as a by-product of nickel mining. Canada's production of cobalt from

Table 4
World copper reserves

	Reserves (million short tons)	Percentage of total world reserves
North America		
United States	93	18%
Canada	34	7%
Other	33	7%
Total	160	32%
South America		
Chile	93	18%
Peru	35	7%
Other	22	4%
Total	150	30%
Africa		
Zaire	28	6%
Zambia	32	6%
Other	10	2%
Total	70	14%
Asia (Total)	30	6%
Oceania (Total)	20	4%
Europe (Total)	7	1%
Communist States	66	13%
World Total	503	100%

the nickel and copper sulfide deposits in Ontario and
Manitoba provided 5% of the noncommunist supply in
1975.[13] Finland supplies 4% of the noncommunist
production.[14] The remaining 15% of the noncommunist
production comes from the nickel deposits of the
western Pacific, including Australia, New Caledonia,
and the Philippines.[15] Indonesia is expected to
become a cobalt producer in the near future with the
development of a nickel-cobalt project at Gag
Island.[16]

The by-product nature of cobalt limits the
potential expansion of production. Only in Zaire,
where cobalt content may range from a high of 2% to
considerably lower values, can the production of
cobalt be relatively independent of the production of
copper. In nickel deposits, the ratio of nickel to
cobalt may average about 15:1 over the entire deposit,
so increases in cobalt production must be accompanied
by considerably larger increases in the mining rate
of nickel. For this reason, the production of cobalt
in most countries is relatively insensitive to
changes in market price.[17] Barring political
closures, however, the production in Zaire can vary
to meet the demand that is not met by other sources.[18]

Production of cobalt during the next decade will
probably be limited to deposits of known economic
potential. The most prominent areas are Zaire and
New Caledonia, but production increases in these
areas may be limited by factors other than geology.
In Zaire, the production of more cobalt would require
new facilities for processing of the copper ore and
recovery of cobalt. The financial state of Zaire,
however, is not encouraging to investors, and further
investment in new facilities may be low. In New
Caledonia the production of cobalt from the nickel
laterites will be limited by technical and economic
factors. First, the ratio of nickel to cobalt in
New Caledonia, depending on the sources of ore,
ranges from 30:1 to 15:1, and further, the smelting
methods that may be used there result in recovery of

less than half of the contained cobalt. Thus cobalt
production is limited to about 3% of the nickel production. Second, even that rate of production is
dependent on the construction of improved processing
facilities and selection of ore for high cobalt content. If the improved facilities are not used,
cobalt production will be even lower. For example,
in 1975 the recovery ratio for New Caledonia ore was
almost 70 tons of nickel per ton of cobalt.[19]
Potentially this ratio could double, resulting in
production of over 4,000 tons of cobalt at current
production levels of nickel. The growth of the
nickel market has slowed in the past few years and
is not expected to pick up again until the middle
1980s, so it is unlikely that cobalt production as a
by-product of nickel will increase other than by
increases in processing efficiency, which are likely
to be small. If future production of cobalt is to
come from known reserves, then Zaire will continue to
be the major supplier.

Manganese
Production of manganese ore is located primarily in
six countries. These countries are, in order of production capacity, the Soviet Union, South Africa,
Gabon, Brazil, Australia, and India. The distribution of world manganese in 1975 production is shown
in table 5. South African production accounts for
37% of the noncommunist production. The production
of Gabon, Zaire, Ghana, and Morocco added to that
of South Africa accounted for 55% of the noncommunist
production in 1975.[20]
     The trends in the development of manganese mines
are shown in table 6. Estimates through 1985
indicate that the Soviet Union will remain the largest
producer, and South Africa will remain the second
largest producer and the major supplier of manganese
to noncommunist nations. The completion of the Trans-
Gabon railway in the early 1980s will allow the Gabon
to increase its production by 36% from 1975 to 1985,

Table 5
World manganese production, 1975

	Production (thousand short tons)	Percentage of total world production
USSR	3,395	31%
South Africa	2,588	24%
Gabon	1,233	11%
Brazil	875	8%
Australia	857	8%
India	624	6%
China	440	4%
Other	798	7%
Total	10,810	100%

Source: U.S. Bureau of Mines.

so it will continue to account for about 10% of world production.[21]

The world reserves of manganese are primarily located in large deposits that are currently being exploited. Unless another large deposit is discovered, or the manganese in deep seabed nodules is exploited, the current producers will continue to supply the world's needs for manganese into the next century.

The size of the world manganese reserves is a point of controversy. In recent testimony in support of the deep seabed mining industry, Mr. Phillip Hawkins, a vice-president of U.S. Steel, cited in-house research showing world reserves of less than 700 million tons, in contrast to U.S. Bureau of Mines/ Geological Survey estimates of two billion tons.[22] U.S. Steel argues that the indicated and inferred resources that surround existing manganese deposits will be more difficult and expensive to mine and process and should not be classed as reserves. For this study, the U.S. Steel estimates of measured reserves are accepted, while the USBM/USGS estimates

Table 6
Percentage distribution of manganese ore production*

	1970	1975	1980	1985
USSR	38%	39%	37%	37%
South Africa	15%	18%	19%	18%
Gabon	8%	10%	10%	11%
Brazil	10%	8%	7%	7%
Australia	4%	7%	7%	6%
China	5%	5%	5%	6%
India	9%	6%	5%	4%
Other	11%	7%	10%	10%

*Because the grade of ore varies with location, this
table does not indicate the distribution of metal
production. This information, for 1975, is given in
table 5.

Source: C. R. Tinsley, "Manganese Gains Depend on
Rate of Recovery in Steel Industry," Engineering and
Mining Journal, Vol. 178, no. 3, p. 99.

of indicated and inferred resources are taken to
represent alternate sources of manganese which may
be mined after the more economic sources are depleted.
     The reserve estimates of the U.S. Steel Corpora-
tion are shown in table 7. The world reserves are
significant in six countries, but two, the Soviet
Union and the Republic of South Africa, account for
82% of the total.

OCEAN RESOURCES
Exploration of the deep ocean has shown that much of
the seabed is covered with manganese nodules. In
some cases, the nodules have been found to have concentrations of nickel, copper, and cobalt sufficiently
large to suggest that they might be economically
exploited.) Currently more than twenty mining and
petroleum companies have invested in research programs to develop the technology to mine and process

Table 7
Proven reserves of manganese

	Reserves (million tons of contained metal)	Percentage of total world reserves
USSR	306	45%
South Africa	249	37%
Gabon	57	8%
Australia	33	5%
Brazil	11	2%
India	3	1%
Other	19	3%
Total	678	100%

Source: U.S. Steel Corporation.

the nodules. (Most of the development has been done by private companies) but information available from academic institutions and from research programs conducted by the Department of Commerce, the Department of the Interior, and the National Academy of Sciences confirms that the nodule resource may be commercially mined for the content of nickel, copper, cobalt, and possibly manganese.

The contribution of a nodule mining operation to the supply of metals depends on the amount of metal contained in the nodules and the rate at which they are mined. Nodules that are recovered in the high-grade province of the northeast equatorial Pacific may contain as much as 1.5% nickel, 1.3% copper, 0.24% cobalt, and 25% manganese, and the average grade is likely to be about 2.3% combined nickel and copper. A single mining operation is expected to recover about 3 million tons of nodules per year, so the annual metal production of a single mine, as shown in table 8, could be as much as 42,000 tons of nickel, 37,000 tons of copper, and 4,000 tons of cobalt. The decision to recover manganese would depend on the market price for the product, but

Table 8
Metal production from a hypothetical first-generation nodule mine[1]

Metal	Nodule content	Recover efficiency	Amount recovered (short tons/year)	Percentage of 1975 world production
Nickel	1.5%	95%	42,000	4.7%
Copper	1.3	95	37,000	0.5
Cobalt	0.24	60	4,000	11.0
Manganese [2]	25	(100)	(750,000)	6,9

1. This calculation is based on a mine that recovers 3 million dry short tons of nodules per year. The content of nickel, copper, and cobalt represents that of the early mines and will be lower for later mining operations.
2. Since the recovery of manganese is not certain, the figures in this table represent the maximum potential production if all manganese is recovered from the ore. The actual production may be zero or an intermediate amount.

theoretically the nodules could supply up to 750,000
tons of manganese per year. In comparison to the
1975 world mine production of these metals, a single
nodule mine could supply about 5% of the nickel production, 0.5-1% of the copper, and 12% of the cobalt.
If all of the manganese were recovered, it would
account for about 7% of the 1975 world production.
        Nodules are found in many parts of the world, but
only in one region of the Pacific Ocean have they
been found in economic grade and abundance. Even in
this region, there is insufficient data for an
accurate assessment of the size of the resource.
Since 1975 there have been at least five attempts to
estimate the extent of the nodule resources. The
results of these estimates are shown in table 9. The
earliest estimates, made by Alan A. Archer, and by
Alexander Holser, were both based on an assumption
that metal content and surface coverage are independent.[23] Thus, it was assumed that areas of acceptable
metal content would have a possibility of sufficient
surface coverage to be a minesite. Recent analysis
of data on nodule deposits indicate that these factors
may be inversely proportional, so that minesites are
less likely than predicted by Archer and Holser. The
remaining three estimates are limited to the area in
the northeast equatorial Pacific where exploration
has confirmed both acceptable grade and abundance.
In this region, the estimates of the number of potential minesites, based on technology that may be
developed during this century, range from 4 to 113,
with Pasho suggesting a mean of 27 and Frazer 28.[24]
If there are 28 minesites, then the reserves of the
area can be estimated. In Frazer's paper the minimum
average grade of the nodule deposits is assumed to be
1.24% nickel and 1.03% copper. If the deposits also
average 0.24% cobalt, then the total reserves of the
area would be about 50% of the current world nickel
reserves, 5% of the world copper reserves, and 350%
of the cobalt reserves. The recovery of manganese
from the nodules is uncertain, but the manganese

Table 9
Metal reserves in the Pacific nodule deposits
(number of potential minesites)

Estimator	Year	Range	Mean value or best guess
Archer	1975	55-166	--
Holser[1]	1976	80-185	--
Pasho	1976	4-113	27
Frazer	1977	14-56	28
Thiry	1977	8-22	--

1. Holser estimates 20 to 25 first-generation nodule
minesites in the Pacific deposits between the Clarion
and Clipperton Fracture Zones, and a maximum of 50
sites in this region using "third-generation" mining
technology.

content of the 28 minesites is equal to 80% of the
U.S. Steel estimate of the world manganese reserves.
The metal content of the minesites is shown in table
10, as well as the ratio of nodule metal content to
the cumulative world demand from 1976 to 2000.
    Table 10 shows that ocean mining has the potential
to become a major source of nickel and cobalt. In
addition, there are several possibilities for the
development of the nodules into a source of manganese.
Ocean Mining Associates claims that they can recover
one million tons of nodules per year and separate the
contained manganese for use in the steel industry.
The consortia led by INCO is developing a process
that will produce a high-manganese slag that can be
processed to a form of high-carbon ferromanganese,
also for use in the steel industry. Kennecott Copper
is investigating processes to recover manganese from
the tailings of their processing plant. Technical

Table 10
Potential contribution of ocean resources*

	Total seabed resources (million tons)	Ratio of resources to cumulative world demand, 1976-2000		
		Low demand	Medium demand	High demand
Nodules	2,100			
Nickel	26.0	1.00	0.88	0.82
Copper	22.5	0.08	0.07	0.06
Cobalt	5.0	4.39	3.67	3.28
Manganese	504.0	1.34	1.27	1.10

*Based on nodule deposits with average metal content
of 1.24% nickel, 1.03% copper, 0.24% cobalt, and
24% manganese.

analysis of the OMA and INCO systems indicates that
the operations are feasible, but the profitability
of each system is dependent on the market price of
the manganese product.[25] Kennecott has not released
any figures on the feasibility of manganese recovery
from nodule tailings, but the National Materials
Advisory Board, in a study of manganese recovery
technology, has provided an analysis of the costs of
a possible system that would produce a high-grade
manganese ore. This analysis estimated a capital
investment of $160 million and an operating cost of
7.7 cents per pound of manganese would be required to
recover manganese from one million tons of nodules
per year. This estimate assumes that the nodule
tailings are available at no cost. These assumptions
would require a market price of about $2.90 per long
ton unit of manganese for the processing company to
receive a 12% rate of return.[26] Since this price is

almost twice the current market price for manganese
ore, it must be assumed that this particular process
is not currently economic. There may be other processes that can recover manganese at a lower cost,
but without access to proprietary information their
profitability cannot be evaluated. Even if other processes are not profitable, the unprocessed tailings
of the three metal processes could constitute a
resource of manganese available in the future. If
the tailings are stored in the United States, then
they would become the highest-grade domestic manganese
deposit and would be available at less than half the
cost of producing manganese from existing domestic
ores.

CONCLUSIONS
Demand through the end of the century can only be
estimated since data are incomplete, and predictive
methods are limited in their ability to portray the
future. Based on an examination of past consumption
patterns and estimates of future demand, arguments
that worldwide demand for copper, nickel, cobalt, and
manganese will grow more slowly than in the past
appear reasonable. The range of demand for 1985 and
2000 forecast by the Bureau of Mines gives a sufficient variety of demand growth expectations for
determining policy for these four metals.

   Copper production is primarily centered in three
regions: the United States, Chile, and the African
copper belt in Zaire and Zambia. Production in Peru,
New Guinea, and other countries supplements these
three major areas. Copper mining in the United States
is a strong industry with large reserves that assure
continued strength. Copper production in Chile has
been reliable, with the single exception of the
political disturbance in the early 1970s. The
existence of a major copper recycling industry and
the susceptibility of copper to substitution in many
of its uses reduce the impact of any temporary fall
in the primary supply.

     Nickel production is concentrated in three nations,
supplemented by a number of smaller producers. Two
suppliers, Canada and Australia, are developed states
with strong ties to the United States. The third
state, New Caledonia, is under strong French influence
and is a dependable source of nickel. All three of
these nations have large reserves and will continue
to be major suppliers for the rest of the century.
The supply from these states may also be supplemented
by production from a number of developing states as
the price of nickel rises, so that cartel action by
the major producers is unlikely to have a major effect
on the supply.
     Cobalt is produced as a by-product of nickel and
copper mining, so its supply cannot be significantly
increased alone. This by-product relationship prevents the use of the large reserves of the metal in
areas such as New Caledonia and gives Zaire a special
importance as a supplier. Because of the high cobalt
content of its copper ore, Zaire is able to supply
over 50% of current world demand. Because revenues
from the sale of copper and cobalt are essential to
the government of Zaire, however, the mines are an
attractive target to forces that wish to weaken or
remove that government. Should this cobalt production be lost, other producers could not make up the
loss, so the price of the metal would rise, resulting
in substitution of other materials where possible.
Applications where cobalt is most valuable, including
some super-alloys, tool steel and cemented carbides,
and some magnetic alloys, could be met by the supply
from non-Zairean sources supplemented by a slight
increase in recycling. An increased supply of cobalt
from the deep seabed may therefore be considered a
question of economic importance, but not one of
national security.
     Mine production of manganese ore is centered in
six countries: Brazil, Gabon, Australia, South
Africa, India, and the Soviet Union. This diversity
compensates for the fact that several states may not

be able to assure undisturbed supply due to political
or economic forces. The distribution of high-grade
reserves, however, indicates that in the future South
Africa and the Soviet Union will be the major producers as the other producers deplete their high-grade
ore and are left with a lower-grade, less-profitable
product. By the end of the century, it is possible
that the South African manganese production may be
subject to political disruptions, either by anti-
government forces or by the government, and the Soviet
Union cannot be relied upon as a source. Since there
are no substitutes for manganese in the steel industry, it is important to explore alternative sources
of supply.

The main contribution of ocean mining to national
security lies, therefore, in its potential to reduce
our dependence on potentially unreliable sources of
manganese. Ocean mining may contribute to this goal
in two ways. First, manganese may be recovered
directly from the nodules in the processing plant.
Such recovery technology has not been used before,
and all new research in this area is held by private
firms, so it is not possible to predict when it can
be implemented.

Second, the manganese-bearing waste from the
processing plant of a nickel, copper, and cobalt
operation can be stored for the future. The technology for processing this waste is available, but
would require a doubling in the price of manganese
ore before manganese from nodules becomes competitive
with land-based ores. The waste from the nodule
plant would become a deposit of manganese to be
exploited at a future date.

REFERENCES

1. Minerals in the U.S. Economy, U.S. Bureau of
Mines, Department of the Interior (Washington, 1977),
p. 56.

2. Ibid.

3. Ibid., p. 25.

4. Ibid.

5. Copper - 1977, U.S. Bureau of Mines, Department
of the Interior (Washington, 1978), p. 4.

6. Minerals Yearbook, U.S. Bureau of Mines, Department of the Interior (Washington, 1973).

7. Economic Impact of Environmental Regulations in
the United States Copper Industry, Arthur D. Little,
Inc. (Cambridge, 1978), p. I-3.

8. Ibid.

9. Copper - 1977, p. 6.

10. Ibid.

11. Ibid.

12. Minerals in the U.S. Economy, p. 23.

13. Ibid.

14. Ibid.

15. Ibid.

16. "1976 Survey of Mine and Plant Expansion,"
Engineering and Mining Journal, Vol. 177, No. 1, p. 80.

17. Cartelization in the World Cobalt Market, Charles
River Associates, Inc. (Cambridge, 1976), p. 1.

18. Ibid., p. 2.

19. Minerals in the U.S. Economy, p. 50.

20. Ibid.

21. C. R. Tinsley, "Manganese - Gains Depend on Rate or Recovery in Steel Industry," Engineering and Mining Journal, Vol. 178, No. 3, pp. 98-104.

22. Robert L. L'Esperance, Statement before the Subcommittee on Oceanography, Committee on Merchant Marine and Fisheries, U.S. House of Representatives, April 19, 1977, p. 249.

23. Jane Z. Frazer, "Manganese Nodule Reserves: An Updated Estimate," Marine Mining, Vol. I, Nos. 1 and 2 (1977), p. 105.

24. Ibid., pp. 119-120.

25. Dames and Moore and EIC Corporation, Description of Manganese Nodule Processing Activities for Environmental Studies, Vol. I (Washington: National Technical Information Service), pp. 2-5 and 2-11.

EFFECTS OF DEEPSEA MINING ON INTERNATIONAL MARKETS FOR COPPER, NICKEL, COBALT, AND MANGANESE

Bernard J. Reddy and Joel P. Clark

Any discussion of the economic and political costs and benefits of ocean mining, and therefore any discussion of possible regulatory systems, must be based on a solid understanding of how ocean mining will affect the markets for the metals involved. These issues (and others) currently are the subject of a study being performed by Charles River Associates and the Department of Materials Science and Engineering at M.I.T. This paper reports some preliminary findings of the ongoing study.

OCEAN MINING OPERATIONS
Most public discussions of potential ocean mining operations have been based on a common set of facts or assumptions. For example, it is generally assumed that a typical ocean mining operation will recover 3 million metric tons (dry weight) per year of nodules grading about 1.25% nickel, 1.0% copper, 0.25% cobalt, and 25% manganese. Three of the four international ocean mining consortia that include U.S. firms are planning operations of this magnitude; one plans to recover only 1 million tons of nodules per year.

One such operation, therefore, could produce annually about 34,000 metric tons of nickel, 27,000 metric tons of copper, 7,000 metric tons of cobalt, and 675,000 metric tons of manganese (see table 1). In 1976, a single ocean mine site could have supplied over 20% of the nickel used in the United States, all of the cobalt, and almost 70% of the manganese, but only 1% of the primary copper. By 1990, however, assuming that demand for each of these metals in the noncommunist world grows at an annual rate of 3%, it will take four such ocean mining operations to account for 20% of the nickel market, 60% of the cobalt market, 30% of the manganese market, and 1% of the copper market.

Table 1
Comparison of ocean mining output and 1976 markets
for nickel, copper, cobalt, and manganese (thousands
of metric tons)

Metal	Single ocean mining operation*	U.S. primary consumption	Noncommunist primary consumption
Nickel	34	148	490
Copper	27	2,600	8,700
Cobalt	7	7	30
Manganese	675	1,000	6,000

*Assumes 3 million tons per year of nodules (grading
1.25% nickel, 1.0% copper, 0.25% cobalt, and 25%
manganese) are processed and 90% of the metals in the
nodules are recovered.

Source: CRA world market models for nickel, copper,
cobalt, and manganese.

MARKET IMPACTS
Since first-generation ocean mining operations are
expected to affect the copper market only minimally,
these impacts will be ignored in this paper. The
effects on the markets for nickel, cobalt, and manganese may be substantial. Analyzing the impacts on
the nickel and cobalt markets is relatively straight-
forward: CRA's forecasting models for these markets
can be used directly to show how the advent of ocean
mining will affect prices, consumption, and production in both developed and less-developed countries.
    To analyze the effects of ocean mining on the
nickel and cobalt markets, three forecasts were
generated for each market for the period 1980-2000.
Each forecast assumed that economic activity, both in
the United States and abroad, would grow at an annual

rate of 3%. The first forecast looked at the future
of the two markets in the absence of ocean mining.
The second assumed that four ocean mining operations--
three recovering 3 million tons per year of nodules
and one recovering 1 million tons--would enter into
production in 1990. In the third forecast, two
additional ocean mining operations, at 3 million tons
each, were assumed to come on-stream in 1995. The
price forecasts for the three scenarios were compared
to assess the likely impacts of first-generation
ocean mining operations on the prices of cobalt and
nickel.

The analysis required for the manganese market is
considerably more complex, for several reasons.
First, it is not clear that manganese from the nodules
can be recovered in a form that currently could compete with the standard ferromanganese used in high-
tonnage steel production. Second, if a high-purity
manganese could be recovered at a price substantially
below the current price of manganese metal, a new
family of manganese-based alloys might be developed
for use in a wide range of applications. Finally,
considerable uncertainty exists about future manganese prices in the absence of ocean mining. Some
industry observers believe that only moderate price
increases (in real terms) will be needed in the future
to encourage the production of manganese from land-
based deposits. Others, however, believe that
prices will have to double or triple over the next 20
years for supply to keep pace with demand. The first
two of these issues will be discussed here; the third
is beyond the scope of this paper.

COBALT
The cobalt market has been in a state of disarray for
the past year, due to production problems in Zaire,
the major noncommunist cobalt producer. As a result,
the producer price soared from about $6.00 per pound
at the beginning of 1978 to $20.00 per pound by the
end of the year. When the current market difficulties

are resolved, the price seems likely to return to a
level of about $10.00 per pound (all prices are in
1978 dollars). The introduction of four ocean
mining operations in 1990, as described above, would
cause an immediate drop in the cobalt price of at
least 25%. Two additional mining operations coming
into production in 1995 would drive the price at
least another 10% below the base level. This would
imply a price of $6.50 per pound, rather than $10.00.

In fact, the sudden influx of such large quantities of cobalt onto the market could lead to a breakdown in the structure of the market, which in turn
could lead to substantially lower prices. Many
market analysts believe that the price of cobalt
normally should be about two to three times the price
of nickel. At a long-run nickel price of $3.00 per
pound, this would imply a normal cobalt price of
$6.00 to $9.00 per pound. The price forecasts described above both fall in this range. However, ocean
mining might drive the price of cobalt down to 1.5
times the nickel price, or about $4.50, in the long
run. This would be a decline of over 50% from the
$10.00 per pound level.

In short, first-generation ocean mining operations
are likely to lead to a decline of at least 25 to 35%
in the price of cobalt, and a drop of 50% is quite
possible. More rapid growth in ocean mining could
lead to even further price declines.

NICKEL
The base forecast for the nickel market showed that
the nickel price is likely to increase from its
current level of about $2.00 per pound to about $2.80
per pound by 1990 and $3.00 by 2000. These forecasts
may be somewhat low because they probably do not fully
reflect the substantial increase in capital costs for
mining that have occurred in the late 1970s. Four
ocean mining operations starting up in 1990 would
lead to a 6% decline in the nickel price, and two
more operations coming on-stream in 1995 would cause

an additional 2% drop. Nickel production in the less developed countries would be affected only slightly-- about a 2% decline.

A more rapidly growing ocean mining industry would certainly cause further decline in the nickel price. However, it seems unlikely that the nickel price could be forced as much as 25% below the base level, or possibly even 15%, because the profitability of ocean mining would be severely affected by such reductions in the price of nickel.

MANGANESE

There is some question as to whether the manganese in nodules can be recovered economically, since a single processing operation of 3 million metric tons of dried nodules would yield approximately 750,000 tons of contained manganese. (In this and the following sections manganese and steel data are expressed in short tons.) If the manganese were to be recovered as relatively pure metal, a 3 million metric ton nodule processing operation would produce about 11 times as much as current world consumption. However, if the manganese were to be priced to compete with standard ferromanganese, the output of that operation would be only about 11% of the current consumption of the world's market economies. Thus, the markets in which the manganese from the nodules would compete appear to be critical. If the manganese is to be recovered to compete as a relatively pure metal, sub- stantial new markets will have to be found before such uses are economically viable. If the manganese can be priced to compete with standard ferromanganese, the existing markets may be adequate to absorb a sub- stantial portion of this new source of supply. This section provides a rough assessment of the potential for utilizing manganese derived from deepsea nodules in various forms: standard ferromanganese, low- carbon ferromanganese, and manganese metal. First, an estimate is made of current consumption patterns of the various forms of manganese in the United

States. (This is the only country for which detailed
statistical information is available at this time.)
Then forecasts of likely technological developments
from 1978 to 1990 are made, and the manganese consumption patterns are altered accordingly. Finally,
the effects of potential price decreases on the consumption of various forms of manganese are assessed.

Current Consumption Patterns and Future Technological Trends

Approximately 95% of the manganese contained in
minerals and ores is consumed in the production of
steel, the remainder going to other alloys, dry-cell
batteries, and chemicals. Apart from manganese
ore--a desirable addition to the blast furnace when
the iron ore is deficient in manganese--the cheapest
form of manganese is standard (high-carbon) ferromanganese, containing about 78% Mn. The most expensive form is nitrided, electrolytic manganese metal,
which is used as an alloying element in some alloy
and stainless steel and in the nonferrous metals
industry.

Manganese is used in steelmaking principally to
counteract the adverse effects of FeS inclusions,
which melt at low temperatures and contribute to
"hot-shortness" problems. (Steels that are "hot-
short" crack when hot-deformed.) The manganese
additions lead to manganese-sulphur compounds that
are either removed in the slag or retained as MnS
inclusions in the steel.

Manganese is also used as an alloying element in
the iron-carbon system to increase the hardenability,
strength, and toughness of the steel. A steel containing manganese, compared to a plain low-carbon
steel, will harden with less severe quenching, harden
to a greater depth, show an improved response to
tempering, and exhibit greater low-temperature
ductility.

Steel Industry

The demand for manganese is closely tied to the production of steel, with the average for all steelmaking
operations at about 14 pounds of manganese per ton of
steel. This figure has declined slightly over the
past 10 years. The major product forms in which manganese is used are low-carbon flat-rolled (sheet and
strip) steels, other carbon steels, alloy steels,
and stainless and heat-resisting steels. Of these
grades, the compositional advantage of a high-purity
manganese product derived from the nodules--i.e.,
negligible carbon and phosphorus contents--would be
of greatest importance in the low-carbon flat-rolled
and stainless steels.

Before estimating the current manganese consumption in each of these categories it is necessary to
examine 1978 steel production figures, shown in table
2. Total raw steel production in the United States
has been depressed in the past few years and has not
yet reached its record level of about 152 million
short tons produced in 1973. Projections are that
future growth will be in the range of 1.5-2.0% per
year in the next 15 years. Data on manganese consumption by use and product appear in table 3.

The average manganese content of the low-carbon
flat-rolled steels is about 0.35%. If the residual
manganese (from pig iron and scrap after oxidation)
is 0.1%, then, with manganese recovery from alloy
additions of 75%, about 0.33% Mn per ton is required.
For the total production of this grade we therefore
require about 200,000 tons of manganese in various
forms. Currently about 10% of this requirement
(varying according to particular steel company practice) is added as electrolytic or very-low-carbon
ferromanganese.

There are two opposing trends to be considered
regarding the use of low-carbon or manganese metal,
however. First, in many of these steels, with carbon

Table 2
U.S. steel production, 1978 (millions of short tons)

Type of Steel	1978 production
Low-Carbon Sheet and Strip	
Hot-rolled	19.6
Cold-rolled, commercial quality	11.4
Cold-rolled, drawn quality	12.6
Tinplate	7.9
Galvanized	8.4
Subtotal	59.9
Other Carbon	56.7
Alloy	18.1
Stainless and Heat-Resisting	2.0
Total	136.7

Source: Annual Statistical Report of the American
Iron and Steel Institute

contents from 0.06 (max.) to 0.23%, the steelmaking
technique is such that, to reach the desired final
carbon content, the oxidation processes must be
carried to such a low carbon level that slag losses
(and hence overall iron yield, heat time, oxygen
consumption, and refractory usage) are not optimum.
This is particularly true in the basic oxygen furnace, which now accounts for more than 60% of the
production. But the carbon level must be lower than
the desired final analysis if standard high-carbon
ferromanganese is used for the final addition. At
present the optimum economic combination requires
that roughly 10% of the final manganese addition be a
low-carbon form, like electrolytic manganese. If the
price of essentially pure metal derived from the
nodules were competitive with standard ferromanganese,
an increased usage of up to 25% of final additions

Table 3
Estimated U.S. consumption of manganese, 1978 and 1990 (short tons)

	Estimated 1978 Consumption			1990 base projections		
	Standard FeMn	Low-C FeMn	Metal	Standard FeMn	Low-C FeMn	Metal
FERROUS						
Low-C flat-rolled sheets	180,000	13,000	7,000	215,000	16,000	8,000
Other C steels	560,000	10,000	1,000	670,000	12,000	1,000
Alloy steels	165,000	13,000	2,000	264,000	21,000	3,000
Stainless and heat-resisting steels	13,000	2,000	3,000	16,000	3,000	4,000
Total ferrous	918,000	38,000	13,000	1,165,000	52,000	16,000
NONFERROUS						
Aluminum alloys	0	0	12,000	0	0	22,000
Other alloys	0	0	3,000	0	0	4,000
Total Nonferrous	0	0	15,000	0	0	26,000

Rates of growth: 15%, C steels; 4%, alloy steels; 2%, stainless steels; 5%, Al alloys; 3%, other nonferrous alloys.

could result; i.e., the market could be increased to
about 50,000 tons.

Second, for the deep drawing qualities (including
some fraction of the galvanized and tinplate steels)
recent developments show that lower manganese contents may be desirable. In particular it has been
shown[1] that a more desirable drawn texture (defined
by the value

$$R = \frac{\text{percent variation in width}}{\text{percent variation in thickness}}$$

and determined by a tensile test on strip samples)
results if the Mn content of the steel is low (less
than 0.15%).

If it is assumed that this development is the way
of the future, it must still be restricted to the
cold-rolled drawn-quality steels and perhaps half of
the combined tinplate and galvanized steels, for a
total of about 20 million tons. With a manganese
reduction from 0.35 to 0.1% Mn (which is almost
exactly the expected residual), the reduction in
manganese additions is 68,000 tons per year, of which
about 10% would be grades utilizing low-carbon or
manganese metal.

The average manganese content for the balance of
the carbon steel production is estimated at 0.95%.
If the residual manganese after oxidation is about
0.2% and we again assume 75% recovery of manganese
from final additions, the total manganese requirement
for this group is about 571,000 tons. However, the
carbon content is such that essentially all of this
would be added as silicomanganese and standard high-
carbon ferromanganese.

For certain of these products--structurals,
plates, and pipe, in particular--it is likely that
the future trend will be high-strength products.
While it is very difficult to estimate, the average
manganese content could increase to 1.1%. For the
present steel production total of about 30 million
tons, this would require an increase of about 59,000
tons of manganese (added as standard ferromanganese).

Such an increase could be offset by other factors. Increasing the manganese content above certain levels may cause difficulties in welding (because of the response of the heat-affected zone), especially in the field. Higher-manganese steels are said to give pouring problems in strand (continuous) casting; the use of continuous casting is expected to increase substantially between 1978 and 1990. These problems might reduce the increased demand for manganese to about 30,000 tons.

Alloy steel production in 1978 totaled about 18.1 million tons. At an average manganese content for the whole group of 1%, the total manganese added (mostly as standard ferromanganese) is about 180,000 tons, taking into account residual manganese and expected recovery rates.

While the alloy steel market is expected to grow at a slightly higher rate than that for carbon steels (perhaps 4% per year) there are other factors, mostly economic, that must be taken into account. A trend toward substitution of low alloy grades for higher alloy grades, wherever possible, is apparent. In turn, we may expect a trend toward substitution of higher-strength carbon steels (with higher manganese contents) for alloy steels. Even an optimistic prediction would suggest an increased demand for manganese of only about 12,000 tons per year. Offsetting factors include, for example, the desire of seamless tube producers to keep manganese content as low as possible because of quench-cracking problems. The increased demand would be for manganese in the form of standard ferromanganese, since none of these steels is a low-carbon variety.

Special high-manganese steels, such as Hadfield steel containing 12-13% Mn and Langley allow containing 20% Mn, might be considered separately. At present the use of manganese for Hadfield steel amounts to about 6,000 tons per year. Since the big application is where wear is a problem, as in railroad crossings and cars, mining, construction and tunneling equipment, any developments in rail

transportation will be significant. There is at
least a potential for a 5,000 ton increase. Manganese metal would, however, seem to offer no
significant advantage over standard ferromanganese.
    Recent trends in the stainless steel industry
have been for the Cr-Ni austenitic steels (2% Mn) to
be replaced by the straight Cr (1% Mn) and Cr-Mn
(7.5% Mn) steels. These trends, if continued, will
simply balance out the manganese demand. The availability of a cheaper source of manganese metal from
the nodules could change the relative trends,
although an optimistic forecast would suggest an
increase of no more than about 1,000 tons of manganese per year.
    Table 4 shows the effects of all the changes
discussed above on manganese consumption in the
United States. It must be emphasized that these
changes, while based on the present annual consumption of about 918,000 tons standard ferromanganese
and silicomanganese and 51,000 tons of low-carbon
and electrolytic manganese, will come about over a
period of years. They are not to be construed as
continuing, annual changes. Table 5 shows the projected consumption in 1990 of various forms of
manganese, taking into account the effects of potential technological developments and the projected
growth rates of the alloy markets.

Potential Effects of Price Reductions
The price elasticities of demand for the various
forms of manganese are shown in table 6. Data for
calculating these elasticities were obtained from
detailed personal and telephone interviews with
representatives of industries consuming manganese.
The objective of these interviews was to determine
how much more manganese would be consumed (if any)
at parametrically decreasing prices. As can be seen
from table 7, a decrease in the price of ferromanganese would not result in substantial increases in
consumption due to technological constraints. Even

Table 4
Effects of potential technological developments on
U.S. manganese consumption

Product and development	Increase or (decrease) in manganese usage (tons per year)	
	Standard FeMn	Low-C electrolytic
FERROUS		
Low-carbon flat-rolled:		
a) Lower carbon and phosphorus requirement	(30,000)	30,000
b) Low-manganese deep-drawing	(61,000)	(7,000)
Other carbon steels:		
a) Higher strength requirements	59,000	
b) Welding and continuous casting problems	(29,000)	
Alloy steels:		
a) Substitution of low alloy and carbon grades	12,000	
b) Hadfield steels	5,000	
Stainless steels:		
a) Cr and Cr-Mn growth		1,000
Total ferrous (if all changes take effect)	(44,000)	24,000
NONFERROUS		
Mn-Al bronze propellors	0	1,000
Mn-Cu damping alloys	0	2,000
Die casting Mn-Zn-Cu alloys	0	1,000
Others	0	1,000
Total nonferrous	0	5,000

Table 5
1990 projected consumption with technological
developments (thousands of short tons)

	Standard FeMn	Low-C FeMn	Metal
Low-C flat-rolled sheets	124	31	16
Other C steels	700	12	1
Alloy steels	281	21	3
Stainless and heat-resisting steels	16	4	4
Nonferrous alloys	0	0	31
Total	1,121	68	55

in the nonferrous industry, there is little
possibility of increased consumption with price
decreased in current applications. Although the
results cannot be quantified at this time, however,
there is a distinct possibility for the development
of new manganese-based alloys in substantial tonnage
applications if the price of relatively pure manganese is decreased 25 to 50% from 1978 levels.
    There are essentially two ways in which manganese
can be recovered from the nodules. One is to recover
the manganese as a slag which could then be stockpiled or substituted for terrestrial manganese ores
in the normal ferromanganese production process. The
second is to recover the manganese in a relatively
pure form along with the copper, nickel, and cobalt.
    The consortium called Ocean Management, Inc.,
headed by International Nickel Ltd. (INCO), is planning to recover manganese by the first route. The
goals of this process as outlined by INCO researchers[2]
are to recover a high percentage of the copper,

Table 6
Effects of price reductions on U.S. manganese
consumption

Price Elasticities of Demand	
	E_d
FERROUS	
Standard FeMn	0.12
Low-C FeMn)	
MnMetal)	0.20
NONFERROUS	
Mn-Al bronze	0.20
Mn-Cu-Al	0.20
Mn-Zn-Cu	0.30
Al-Mn	0.15
Mn-based alloys	?
Weighted average,	
all nonferrous	0.20

E_d = price elasticity of demand.

nickel, and cobalt by a smelting process, rejecting
manganese at an early stage. The manganese would
then be further processed into a standard ferromanganese product. This process, which would recover
on the order of 600,000 tons of contained manganese,
could possibly offer ferromanganese at a reduced
price relative to land-based resources, depending
on the accounting costs attached to the manganese.
However, it is more likely that the price will not be
appreciably different from presently produced ferromanganese since it is not a feed-price-dominated
production cost.
     The more interesting possibility is for the
manganese to be recovered as a relatively pure metal,
possibly containing some silicon but no carbon or
iron. If the manganese were priced at the level of
ferromanganese, it would first saturate the markets

Table 7
Projected effects of price reductions on U.S.
manganese consumption, 1990 (short tons)

	% decrease in price				
	10	20	30	40	50
Standard FeMn	1,134,000	–	–	–	–
Low-C FeMn	69,000	–	–	–	–
Mn metal	56,000	–	–	–	–
Nonferrous (excluding Mn-based alloys)	31,600	32,200	32,900	33,500	34,100
Total	1,290,600				1,293,000

for low-carbon and pure manganese, and the residual
would be left to compete with ferromanganese. The
projected market for relatively pure manganese in the
United States in 1950 is approximately 123,000 to
159,000 tons. For the world the market is about 1.2
million tons. Thus it appears that one or two large-
scale operations recovering relatively pure manganese
could be viable in 1990, depending on the price at
which the manganese would be offered.

In order for larger quantities of pure manganese
to be consumed, new manganese-based alloys would have
to be developed. Preliminary analysis indicates that
the gamma phase of manganese, if appropriately
stabilized, can be used in a range of interesting
alloys.

REFERENCES

1. H. Hu and S. R. Goodman, "Effect of Manganese on
the Annealing Texture and Strain Ratio of Low-Carbon
Steels," Metallurgical Transactions 1 (1970): 3057.

2. R. Sridhar, W. Jones, and J. Warner, "Extraction of Copper, Nickel and Cobalt from Sea Nodules," Journal of Metals (April 1976): 32.

THE NET VALUE OF MANGANESE NODULES TO U.S. INTERESTS, WITH SPECIAL REFERENCE TO MARKET EFFECTS AND NATIONAL SECURITY

James C. Burrows

All four of the commodities which would be produced in significant amounts from manganese nodules--manganese, cobalt, nickel, and copper--can be considered to be critical and strategic minerals. However, the degree to which each is critical varies greatly from commodity to commodity.

To provide some perspective on the major manganese nodule commodities, table 1 presents salient data for 1977 on world consumption, U.S. consumption, production, and imports; and production from a "typical" ocean mining operation for each commodity. The principal implications of these data can be summarized as follows:

I. The United States is very dependent on imports of three of the four commodities--manganese, cobalt, and nickel. Indeed, the United States imports virtually all of its primary consumption of these materials.

II. A small number of ocean mining consortia can account for a significant percentage of U.S. and world consumption of these three materials. Four typical ocean mining operations could account for over 30% of 1990 noncommunist world consumption of manganese (assuming the manganese is recovered and sold for carbon steelmaking), nearly 50% of world consumption of cobalt, and about 16% of world consumption of nickel.

At the risk of considerable oversimplification, the major sources of net benefits to the U.S. economy and U.S. welfare in general of ocean mining are as follows:

1. Profits to U.S. Firms Engaged in Ocean Mining. As these profits are not likely to exceed substantially the competitive return required to attract investment

Table 1
Use statistics for copper, nickel, cobalt, and manganese, United States and total noncommunist world (quantities in thousands of short tons; values in millions of 1977 dollars)

	Copper		Nickel		Cobalt		Manganese	
	Quantity	Value	Quantity	Value	Quantity	Value	Quantity	Value
Mine Production, 1977								
U.S.	1,518		14		0.0		0[1]	
Total noncommunist	8,834		878		16.3		10,920[1]	
Consumption, 1977								
U.S.	2,185	2,230[2]	155	650[3]	8.3	92.6[4]	1,040[1]	232[5]
Total noncommunist	7,535		461		30.0[6]		6,000[1]	
U.S. Imports for Consumption, 1977								
Amount	517	530	167	750	8.8	95.6	1,067[1]	238
As percent of consumption	24%		91%[7,8]		100%[8]		100%[8]	
Projected Consumption, 1990[9]								
U.S.	4,050[6]		283[6]		29.0[6]		1,620[6]	
Total noncommunist	14,930[6]		763[6]		63.8[6]		9,348[6]	
Production, One Ocean Mining Operation	30		30		7.5		750	

Source: U.S. Bureau of Mines, Mineral Industry Surveys; World Metal Statistics, Oct. 1978; Metals Week, annual prices.

1. Figures for 1976. Data for 1977 not yet available.
2. Calculated using average producer price of 51¢/lb.
3. Calculated using average New York dealer price of $2.08/lb.
4. Calculated using average producer price of $5.58/lb.
5. Calculated using average value of imports of $223/short ton.
6. Estimated by Charles River Associates.
7. Calculated as: [1-(U.S. production)/(U.S. consumption)].
8. Additional imports (above consumption) assumed to go into stocks.
9. Without ocean mining.

capital in a risky enterprise, we assume that the
external benefits to the economy of these profits
are small relative to the net value of other external
benefits, and we shall therefore not examine these
profits further in this paper.

2. Lower Market Prices. As discussed below,
these benefits are likely to be substantial for
nickel and cobalt and may be substantial for manganese
if it is recovered and used for producing carbon steel.
As copper production from ocean mining will be a small
percentage of world copper supplies, the impact of
ocean mining on copper prices will probably not be
substantial.

3. Likelihood and Severity of Potential Supply
Disruptions. Supply disruptions in basic materials
can generate enormous economic costs, as evidenced
by the 1973-1974 oil crisis. Ocean mining creates
an alternative source of supply for four strategic
materials. These supplies are likely both to reduce
the probabilities of consciously motivated supply
disruptions and to reduce the impacts of possible
supply disruptions that could occur in the four
materials markets. As discussed below, the expected
value of this benefit may be quite high for cobalt
and manganese, may be significant for nickel, and is
probably fairly low for copper.

4. Reduced Probability and Severity of
Cartelization. As indicated by the success of OPEC
and the International Bauxite Association, the costs
of successful cartelization of a resource can be
substantial. Ocean mining supplies can reduce both
the likelihood of successful cartelization and the
extent to which cartels can increase prices. As
discussed below, the expected value of this benefit
may be substantial for manganese, but is probably not
substantial for cobalt, nickel, and copper.

5. Reduced Risk of Depletion of Reserves. As
the resource base of a mineral is depleted, its
market price should rise over time. Ocean mining
supplies will extend world reserves of the four

minerals involved and should therefore lower the path
of prices over time. The economic benefits of these
lower prices in principle are already accounted for
in the second benefit listed above. However, there
is considerable uncertainty regarding known and
unknown world resources of manganese and cobalt.
Only a limited number of large commercial deposits of
manganese and cobalt minable at prices near current
levels are known to exist, and it is quite possible
that market prices will rise above our forecasts by
the 1990-2000 period. In this case, the economic
benefits of ocean mining may be substantially larger
than implied by some of the numbers presented later
in this paper.
       6. Increased Military and "Political" Security.
Since ocean mining sites will be difficult to defend
in the event of a wartime emergency, we assume that
military security benefits will be minimal. "Political" security benefits (such as prevention of
political blackmail by unfriendly producing countries)
may be significant. These benefits are assumed to be
included in the net benefits of reduced likelihood
and severity of potential supply disruptions.

       In the remainder of this paper, we examine in
somewhat greater depth the net benefits of ocean
mining resulting from lower market prices, reduced
severity of supply restrictions, and reduced
exposure to cartelization. The material presented
here is based on work in progress at CRA under an
NSF grant and is therefore neither complete nor
definitive at present.

NET BENEFITS TO THE UNITED STATES FROM LOWER MARKET
PRICES
As shown by the data in table 1, four "typical"
ocean mining projects in 1990 will account for over
30% of world manganese consumption (assuming the
manganese is recovered and used in the production of
carbon steel), about 16% of world nickel consumption,

nearly 50% of world cobalt consumption, and less than
1% of world copper consumption. It seems clear that
the emergence of ocean mining on even a moderate
scale will lead to declines in the prices of at least
some of the materials involved.

Clearly, consumers benefit from lower prices for
materials purchased: not only will they be able to
purchase the same quantities as before at a lower
total cost, but the lower price will allow them to
purchase greater quantities. Just as clearly,
domestic producers of land-based resources will be
worse off: not only will they receive less per unit
of product sold, but they will not be able to produce as much at the lower prices.

Figure 1 shows in a simplified fashion how these
costs and benefits can be measured for the United
States. This methodology can also be used to assess
economic cost and benefits to other countries,
regions, or the world as a whole. The figure assumes
a hypothetical market in which the United States is a
net importer prior to ocean mining. Before the
introduction of supplies from ocean mining, the
market price is P^2, D^2 tons are consumed, Q^2 tons are
produced domestically, and D^2-Q^2 tons are imported.
If ocean mining reduces the price to P^1, consumption
will increase to D^1, production will decline to Q^1,
and imports will increase to D^1-Q^1. A large percentage of these imports may come from the deep seabed rather than from other countries.

It can be shown that the total gain to consumers
as a result of the lower prices are equal to the sum
of areas A, B, C, D, and E--the rectangle composed of
areas A, B, C, and D represents the lower payment for
the D^2 tons consumed prior to the reduction in prices,
and the area E represents the net gain to consumers
from their increased consumption at the lower price.

It can also be shown that producers lose profits
equal to areas A and B. Area A represents the
reduced payment for the amounts supplied in the new
equilibrium (Q^1), and area B represents the surplus

PRICE

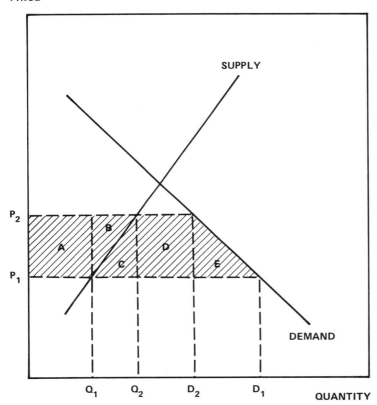

Figure 1.   The economic value of manganese nodule
supplies.

of revenues over resource costs for the sales foregone
as a result of the lower prices (Q^2-Q^1).

Because domestic primary supplies are small for
manganese, cobalt, and nickel, in this paper we assume
that the net benefits to the U.S. economy of lower
prices from ocean mining equal the sum of areas A, B,
C, D, and E for these commodities.

In an ongoing project undertaken jointly by CRA
and M.I.T. for the National Science Foundation, we
are developing estimates of the effects on market
prices of ocean mining. These price effects were
discussed in the paper by Bernard Reddy and Joel
Clark. Table 2 summarizes our interim assessment of
the market effects for nickel and cobalt of ocean
mining and develops estimates of the net benefits to
consumers arising from the lower prices.

Our interim results can be summarized briefly as
follows:

Manganese
It is unclear at present whether and to what extent
manganese will be recovered from manganese nodules
and sold as an alloying material for carbon steel,
by far its largest application. Substantial manganese supplies from even a few ocean mining projects
would swamp the manganese ore market and would almost
certainly lead to substantially lower prices. We
are currently evaluating the likelihood and potential
impacts of substantial manganese being recovered and
sold from nodules, but estimates of market effects
are not yet available. As mentioned above, if market
prices of manganese begin to increase substantially
over time, the economic benefits of ocean mining of
manganese may become very substantial. For example,
there are some estimates that new land-based supplies
of manganese will in the future cost two to three
times as much as existing supplies. If ocean supplies
of manganese prevent the price from rising above
current prices by a factor of 2, the current benefits
to the U.S. economy would be nearly $250 million per

Table 2
Market impacts of deep ocean mining, 1990 and 2000, nickel and cobalt

	Price without ocean mining[1] (1)	Price with ocean mining[1] (2)	Price effect of ocean mining[1] (3)	Consumption without ocean mining[2] (4)	Consumption with ocean mining[2] (5)	Consumption effect of ocean mining[2] (6)	Net benefits to U.S. consumers from ocean mining[3] (7)
Nickel							
1990							
U.S.	2.84	2.63	0.21	311	315	4	131
Total noncommunist	2.84	2.63	0.21	839	1021	182	391
2000							
U.S.	3.04	2.93	0.11	431	439	8	96
Total noncommunist	3.04	2.93	0.11	1129	1240	121	262
Cobalt							
1990							
U.S.	10.00	7.56	−2.44	14.5	15.0	0.5	72
Total noncommunist	10.00	7.56	−2.44	46.4	48.9	2.5	233
2000							
U.S.	10.10	8.14	−1.96	21.3	23.7	2.4	88
Total noncommunist	10.10	8.14	−1.96	64.0	76.0	12.0	275

Column (3) = (2)−(1); Column (6) = (5)−(4); Column (7) = (3)x[(4)+0.5x(6)]x2,000.
Source: Charles River Associates' econometric models of the nickel and cobalt industries.
1.   1978 dollars per pound.
2.   Thousands of short tons.
3.   Millions of 1978 dollars.

year (1978 dollars). After large-scale introduction
of ocean mining, this benefit will grow by 2-3% per
year. Using a conservative capitalization factor of
10, the present value of this benefit as of 1990
(assuming that large-scale nodule recovery has commenced by that date) could be as high as $2.5 billion.
This benefit is uncertain, however. It is not known
how much, if any, manganese will be recovered for
steel usage, nor at what cost it will be recovered.
Furthermore, there is considerable uncertainty over
the future path of manganese prices in the absence of
ocean mining.

Nickel
Our model of the nickel market predicts that nickel
supplies from four nodules projects by 1990 would
depress the nickel price by about 7% and would yield
an annual benefit to U.S. consumers of about $130
million dollars. Using a capitalization factor of
10, the 1990 present value of this economic benefit
is about $1.3 billion (in 1978 dollars).

Cobalt
The cobalt market is currently [January 1979] in a
state of chaos as a result of the interruption in
supplies from Zaire. The producer price has increased
from $6.40 per pound in 1977 to $20 per pound and
open market prices have risen as high as $50 per
pound. In table 2 we assume that by 1990 the market
will have returned to "normal" and that the price
will have declined substantially. This clearly is
a debatable assumption--indeed, CRA is currently
initiating an in-depth study of this very question.
The results in table 2 should therefore be viewed as
indicative and not as definitive. With this qualifi-
cation in mind, table 2 shows that four ocean mining
projects will lead to a decline in cobalt prices of
about 25% and an annual economic benefit to U.S. consumers of $70 million. The 1990 present value of
this benefit as of 1990 is about $700 million (in
1978 dollars).

Copper

Supplies of copper from manganese nodules are not
likely to account for a very high percentage of world
copper supplies. In addition, our analysis suggests
that the long-run supply elasticity of copper in a
price range of $1.00 to $1.25 per pound in 1978
dollars is very high. As a result, the impacts of
ocean mining on copper prices will be fairly small;
no attempt to measure these impacts is made in
table 2.

NET BENEFITS TO THE UNITED STATES FROM THE REDUCED
LIKELIHOOD AND SEVERITY OF SUPPLY RESTRICTIONS

The current turmoil in the cobalt market and the
traumatic events of 1973-1974 in the oil market
(which may be repeated in 1979-1980) are graphic
reminders of the enormous economic costs that can be
incurred as the result of supply disruptions. Supplies of critical materials from ocean mining can
substantially reduce the expected value of the costs
of such disruptions by reducing both the duration and
the severity of a disruption. They can also substantially reduce the costs of preparing for such disruptions through such measures as stockpiling. These
benefits are particularly important for manganese and
cobalt. Ocean mining supplies of coablt, for
example, would probably not have prevented the 1978-
1979 crisis, but such supplies in the future could
deter intentional supply restrictions, whether
economically or politically motivated.

Supplies of manganese from ocean mining could
have equally important effects on the manganese
market. The United States depends on imports for
its primary supply of manganese from high-grade ores,
an essential ingredient in steelmaking. Production
is concentrated, with the top six countries--the
Soviet Union, South Africa, Brazil, Gabon, Australia,
and India--accounting for approximately 90% of world
supplies. The presence of such producers as South
Africa in this list does not inspire confidence in
the security of our manganese supplies. The

availablility of supplies from other sources,
particularly on short notice, is problematical. The
nature of the usage of manganese as an essential
alloying element for steel suggests that the economic
impacts of a severe reduction in supplies could be
catastrophic.

Just four ocean mining sites could supply over
30% of noncommunist world consumption of manganese
from high-grade ores by 1990. The presence of these
supplies would certainly reduce the severity of a
manganese supply disruption and would just as certainly reduce the likelihood that a few major
suppliers might consciously precipitate a crisis for
economic or political gain.

It should be noted that these favorable benefits
might be achieved even if the manganese is not
recovered and sold as an alloying element for steelmaking, if the manganese could be recovered and used
as a steel additive during an emergency. This would
be particularly true if the manganese-containing
tailings had been stockpiled prior to the emergency,
although it appears that this option may not be
technically feasible. At any rate it seems reasonable
to assume that the increased quantities of manganese
potentially available from manganese nodules would
significantly reduce the potential economic cost
from possible manganese supply interruptions.

The contingency benefits of ocean mining for
nickel and copper are probably less important than
for manganese and cobalt. Nickel supplies are more
widely dispersed than manganese or cobalt suppliers,
and Canada is still the world's largest supplier of
nickel, accounting for 47% of noncommunist world
supplies in 1976. Nevertheless, the dangers of
supply interruptions cannot be totally discounted.
Nickel demand is relatively price-inelastic in the
short run; nickel supply is also price-inelastic in
the short run when capacity is fully utilized, so
the consequences of supply disruptions could be
severe.

The contingency benefits for the United States of ocean mining for copper do not appear to be substantial. First, ocean mining supplies of copper probably will not represent a significant percentage of world supplies. Second, the United States is nearly self-sufficient in copper. Third, world copper supplies are fairly dispersed. Finally, the price inelasticities of both demand and of supply appear to be moderate at least. As a result of these factors, a disruption in any one country is not likely to have a pronounced effect on the market--and even if it did the United States, as the result of the importance of U.S. production, would be less severely affected than other consumers.

An earlier CRA study examined in detail the potential costs of supply disruptions in a number of major minerals, including manganese, cobalt, and copper. Determining the expected present value of the costs of potential disruptions is difficult and requires knowledge of such subjective parameters as the probabilities of disruptions. In its earlier studies CRA developed an analytical model to determine these expected present risks of supply disruptions and to develop estimates of optimal contingency stockpiles.

A detailed exposition of this analysis is beyond the scope of this paper, but some of the basic concepts can be illustrated by reference to figure 1. Assume that before the disruption the price is P^1, U.S. consumption is D^1, U.S. production is Q^1, and U.S. imports are D^1-Q^1, and that during the disruption the price rises to P^2, U.S. consumption declines to D^2, U.S. supply increases to Q^2, and imports decline to D^2-Q^2. By analogy to our earlier analysis, the net economic cost of the disruption is equal to the areas C + D + E (the analysis is precisely the reverse of the earlier analysis in which the benefits of increased supplies and lower prices were evaluated). Of course, since adjustments are difficult if the emergency occurs suddenly and without warning, the

supply and demand schedules are likely to be quite
steep (price-inelastic) and the resulting economic
costs higher than they would be with more elastic
schedules.

Developing estimates of the present value of
expected costs of supply disruptions requires knowledge of the following paramaters (among others):

1. The probability of the occurrence of a disruption
in any given time period (inception probability).

2. The probability of a disruption continuing for at
least one additional time period once it has begun
(continuation probability).

3. Demand, supply, and prices in the absence of a
disruption in each year of the forecast period.

4. World market price during each period of the
disruption.

5. Price elasticities of demand and supply during
each year of a disruption.

6. Discount rate.

The CRA optimal stockpile model has been applied
to manganese and cobalt, but not to nickel and copper.
For manganese, the expected present value of the cost
of a disruption under fairly moderate assumptions
was about 2 billion dollars in 1976 dollars (nearly
2.5 billion dollars in 1978 dollars). The cost of
the current cobalt crisis if it continues for a full
year will be over $250 million, depending on the
assumptions made about the prices paid by U.S. consumers. (U.S. consumption in 1977 was about 18
million pounds. The producer price has increased by
about $19 per pound. If consumption during the
crisis remains unchanged and all U.S. consumers pay
the producer price, as opposed to the much higher

merchant price, the higher payments by the U.S.
consumer will total $340 million per year.)

The optimal stockpiles implied by our earlier
analysis are 12 to 18 months for manganese (about
$150 to $225 million at current prices), about 12
months for cobalt (about $200 million at a price of
$10 per pound), and one month for copper (about $500
to $700 million at $1 per pound). The actual strategic stockpiles of cobalt and manganese are considerably larger than these figures. Some stockpiles
might still be warranted even with ocean mining, but
their optimal size would be much smaller than without
ocean mining.

It can be argued that supplies of manganese and
cobalt from manganese nodules could substantially
reduce the expected present value of the costs of
disruptions and might eliminate the need for substantial stockpiles. The potential economic benefits
to the United States are therefore at least $500
million (the costs of economic stockpiles which would
not be necessary) and may be as high as 3-4 billion
dollars (the discounted costs of possible future disruptions in manganese and cobalt, weighted by the
probabilities of the disruptions). These numbers are
highly subjective; nevertheless, they do indicate the
large magnitudes of economic benefits not entirely
achievable.

NET BENEFITS TO THE UNITED STATES FROM REDUCED
PROBABILITY AND SEVERITY OF CARTELIZATION
The success of OPEC and the IBA demonstrates the
potentially sizable economic costs of successful
cartelization of mineral commodities. Since 1973
active efforts have been made to cartelize a number
of mineral markets, including copper.

The potential economic costs from cartelization
appear to be high for manganese but are probably not
significant for copper, nickel, and cobalt. Cartelization attempts have not been successful for copper,
and an earlier CRA study concluded that the likelihood

and impacts to the U.S. economy of successful
cartelization of copper are low. Furthermore,
potential supplies of copper from nodules are not
large enough to have a significant impact on the
copper market. Nickel resources are fairly dispersed,
and Canada would have to be an active supporter for a
nickel cartel to be successful. This does not appear
likely at present. Finally, the cobalt market is
already in effect a monopoly by virtue of Zaire's
dominant position in the market. The impacts of
ocean mining on the market price of cobalt, which in
normal times is set by Zaire, have already been discussed above.

The circumstances in manganese are quite different.
At least three of the five dominant suppliers to the
noncommunist market--South Africa, Gabon, and India--
seem to be good candidates for explicit or implicit
cooperation in a cartel. Alternative supplies at
prices near current levels ($1.50 per long ton unit)
seem limited; it is felt in at least some quarters
that to increase supplies substantially from new
producing areas will require prices two to three
times higher than current prices. If manganese
supplies from ocean mining prevent prices from rising
from current levels, the net economic benefit to the
United States would equal the probability of successful cartelization times the annual net savings of the
reduced price ($250 to $375 million by 1990). If
the probability of a successful cartel is 0.3, this
annual benefit is $75 to $112.5 million. Using a
capitalization factor of 10 the present value of this
benefit as of 1990 would be $750 million to $1.1
billion (already included in the section above as
lower market prices).

SUMMARY
Our research has not progressed to the point where
we can provide definitive analysis of potential
ranges in the net economic value of manganese nodules
to the United States; definitive estimates, at any

rate, are subjective because they require knowledge
of the probabilities of supply disruption and other
similar events. However, the benefits are clearly
substantial. The present value of net benefits as of
1990 are of lower market prices (in the absence of
supply interruptions or cartelization) seems to be
on the order of $2-4.5 billion, depending on the
assumption made about manganese. The present value
as of 1990 of reduced likelihood and severity of
potential supply interruptions appears to be between
$500 million and $3-4 billion. Finally, the present
value as of 1990 of the reduced likelihood and
severity of cartelization may be as high as $1.1
billion. The sum total of these benefits (as of
1990) is on the order of $2.5-85 billion.

These results of course do not provide a complete
picture of the total impact of manganese nodule production. For example, in this paper we have only
examined the net benefits to the United States. The
net benefits to foreign consumers will be even more
substantial than the net benefits to U.S. consumers.
On the other hand, foreign land-based producers will
be adversely affected by ocean mining. Finally, the
political effects of ocean mining are likely to be
far-ranging and difficult or impossible to evaluate.

THE POTENTIAL FOR INCREASED USE OF MANGANESE IN THE METAL INDUSTRY

Nicholas J. Grant

Until about 10 to 15 years ago, manganese was of interest from the point of view of decreasing its use and increasing national self-sufficiency by learning to recycle steel scrap more effectively and to recover manganese from slags and mill scale.

Now our interests are stirred by a potential for significantly increased usage of manganese, including the development of manganese-base alloys. Obviously this change has been induced by the possibility that manganese may become available not only as a high-grade ferromanganese, but also as a pure metal to serve the double purposes of alloying and providing an Mn-base alloy system. This potential is based on the known existence of large quantities of deepsea manganese nodules from which not only Mn, but Ni, Co, Cu, and possibly Mo can be recovered economically.

CURRENT MANGANESE USES IN ALLOYS

Mn is an essential alloying element in steel. Its primary function is to eliminate embrittlement due to the sulfur that is normally present in quantities ranging up to about 0.05%. The Mn is added in amounts 10 to 15 times that of the sulfur to guarantee formation of MnS, which has a high melting point and a discrete spheroidal form of distribution, in contrast to FeS, which has a low melting point and forms thin films at the grain boundary. Mn also helps to provide strength, toughness, and ductility at low temperatures. Alloying additions are normally at the 0.4-0.8% level, but may be as high as 2%.

These two functions in steelmaking account for almost 90% of current Mn usage, or more specifically for almost 900,000 T/yr in the United States.

Mn is also an important alloying element in non-ferrous alloys, most commonly in amounts between about 0.3 and 1%. It is common in aluminum alloys,

but is also used in similar percentages in some
copper-base alloys. The amounts of Mn used for
nonferrous alloying are quite small. Even including
some high-Mn Cu alloys (naval brasses) and some high-
Mn ferrous alloys (8-20% Mn), manganese requirements
here are less than about 100,000 T/yr in the United
States.

FORMS OF MANGANESE
The Mn added to steels at low Mn levels (less than
1 or 2%) is the cheap, high-carbon ferromanganese
(Fe-Mn) whose Mn content varies from about 65 to 70%.
For much higher Mn additions in special steels (3-20%
Mn), the tendency is to add either low-carbon Fe-Mn
or Mn metal (generally electrolytic). In nonferrous
alloys (Al, Cu), where iron is considered to be an
undesirable impurity element, electrolytic Mn must be
used.

The consumption of standard (high-carbon) Fe-Mn
in the United States is about 900,000 T/yr. Consumption of low-carbon Fe-Mn, lump metallic Mn, and
electrolytic Mn is about 100,000 T/yr, of which
electrolytic Mn accounts for about 20,000 T/yr.
Total consumption is therefore about 1,000,000 T/yr
of manganese.

MANGANESE METAL
Almost none of the one million tons of Mn goes into
the production of alloys based on Mn (greater than
50% Mn). In general known Mn alloys contain not more
than about 30% Mn; rare alloys contain 50% Mn. Two
factors are responsible for this situation.

1. Manganese has four crystallographic forms,
which is highly unusual for a metal. Only one of
these forms, gamma, is a ductile crystalline material;
in Mn it exists at high temperatures. It is, of
course, possible to alloy the Mn with other face-
centered cubic (gamma) metals (Cu, Ni, Al, high-
temperature Fe) to stabilize the ductile gamma

structure over a wide range of temperatures and
compositions. To optimize useful properties in
alloys requires a significant research effort. Useful work has already been done to confirm the potential for preserving and using the gamma Mn structure,
but the effort has been small, widely scattered, and
not at all integrated into a well-planned alloy
development program.

2. Pure Mn today is a relatively expensive
metal; its use as an alloying element places it in
competition with less-inexpensive metals. Its use
as a base for Mn alloys will be determined by its
competitive price.

Mn in standard Fe-Mn costs about 28 cents per
pound of contained Mn. Electrolytic Mn costs about
58 cents per pound, commercial aluminum about 55
cents per pound, and copper about 70 (65 to 110)
cents per pound. Iron, by comparison, is about 14
cents per pound.

Obviously, if pure Mn could be produced at a
price nearer 28 cents per pound, one could make a
case for Mn-base alloys at a cost second only to our
steels. Will Mn nodules help to realize that
potential?

Speculation on the subject suggests that our
needs for Co, Ni, and Cu will bring pressure to bear
in the mining of Mn nodules. Quantities large
enough to provide 50% or more of our cobalt needs
would require 1-3 million tons of nodules and would
provide 250,000 to 750,000 tons of Mn. But are
there uses for this manganese? And will the ultimate
charges for Mn production encourage Mn usage or kill
it?

There have been numerous surveys, calculations,
and test programs to answer these and more complicated
questions. The technical aspects look more straight-
forward than do the economic studies. There are com-
plications other than the technical and economic ones,
such as disposal of the high-Mn, high-Fe gangue if

only Ni, Co, and Cu are recovered. There are
unknowns regarding the allocation of costs to the
winning of Ni, Co, and Cu to enhance the price of
Mn. Clearly, technically, there is great and urgent
need to examine more broadly the chemical metallurgy
of the winning of the four or five elements of
interest to us, with emphasis on the desire to make
the Mn outlook more attractive.

SOME HIGH-Mn ALLOYING APPLICATIONS
Skipping the use of Mn in Hadfield steels (6-12% Mn),
in austenitic high-Mn stainless steels (200 series
alloys with 6-20% Mn), and in the high-damping Mn-Cu
alloys (40-80% Mn), because these applications are
now reasonably well known, we shall concentrate on
lesser-known applications where high-Mn alloy content
is meritorious. These applications (along with the
three listed above) offer the most important outlets
for extensive Mn metal usage.

Table 1 shows that even very simple Fe-Mn alloys
form stable structures with extremely attractive
strength properties. High-Mn alloys are also of great
value because of their low coefficient of expansion
and very high resistivity. Usage is not yet large,
but the properties are potentially important.

Table 2 shows some limited data for the simple
Fe-Cr-Mn steels with Mn contents of 25-40%. Room-
temperature strength values are excellent with high
levels of ductility (and toughness). Much more
interesting are the cryogenic test results. Along
with large increases in yield and ultimate strength
at -180°C, ductility values remain very high and
impact properties are excellent; welding performance
is indicated to be good.

These steels should be competitive with the
expensive Fe-9% Ni cryogenic steels if the Mn metal
costs can be kept at the low side of indicated
values (Ni price is about $2.50 per lb versus 28 to
58 cents per pound for Mn).

Table 1

A. Iron–Manganese Alloys with more than 25% Mn

    a) 100% gamma phase (fcc) when slow-cooled from 1600°F.
    b) No transformation of gamma due to cold work.
    c) Stable over a wide temperature range.

For 26 to 29% Mn	UTS, ksi	YS, ksi	Elong, %
Normalize at 1600°F	95.1	23.2	25.0
Normalize plus 60% cold work*	214.0	204.0	2.0

*Selection of lower amounts of cold work would result in a better combination of strength and ductility.

B. Low-Coefficient-of-Expansion Alloys

    50% Mn – 50% Fe $= 10 \times 10^6$ per °C
    50% Mn – 45% Fe – 5% Al $= 9.67 \times 10^6$ per °C
    50% Mn – 45% Fe – 5% Ni $= 8.93 \times 10^6$ per °C

C. High-Electrical-Resistivity Alloys

    70% Mn – 30% Ni 1800×10^{-6} ohm cm
    45% Mn – 45% Ni – 10% Fe 1600×10^{-6} ohm cm

Table 3 shows properties for a high-Mn-Al-C stainless steel free of both Ni and Cr. Mechanical properties at 20°C are significantly better than those for standard 304 and 347 Fe-Ni-Cr stainless steels, with excellent ductility. At 650°C the strength of alloy A is even greater than that of alloys 304 and 247. Note the very high ultimate strength of the B alloy at 650°C.

The tables clearly show that there are excellent properties to be achieved, even in rather simple alloy compositions. More complex alloys simply have

Table 2
Some properties of Fe-Cr-Mn alloys

a) Nominal alloy compositions include (in weight percent):

C	.05	–	.07	CU	.04	–	.05	Cr	0	–	15
P	.004	–	.009	Ni	.02	–	.10	Mn	0	–	40
S	.008	–	.013	N	.025	–	.065	Fe	bal.

b) Structures or selected compositional ranges:

Mn   28% and Cr   15%   Gamma (to 40% Mn)

Mn = 18-28% and Cr   15% =   +   (martensite)

Composition	Structure	Y.S. ksi		UTS, ksi		Elong, %	
		20°C	-180°C	20°C	-180°C	20°C	-180°C
25Mn – 5Cr	+	35	56.5	99	156	62	52
30Mn – 5Cr		35	70	84	148	85	57
50Mn – 5Cr		35	85	73	141	60	55

Note: Impact values at 20°C and -196°C for alloys with greater than 20% Mn are excellent.

Table 3
Some properties of Fe-Mn-Al-C stainless steels (free
of Ni and Cr)

Typical compositions (in weight percent)

	C	Mn	Al	Other	Fe
A	0.76	34.4	10.2	–	bal.
B	1.0	30.0	8.0	1.5 Si	bal.

Alloy	Y.S., ksi	UTS, ksi	Red Area, %	Elong., %
At 20°C				
A	55	108	72	72
304	33	84	75	63
347	39	91	72	50
At 650°C (1200°F)				
A	30	60	32	33
304	12	39	53	38
347	25	45	67	43
B	--	100	--	--

Note: Alloy A with 60% cold work plus 1 hour age at
1290°F (700°C) shows an ultimate strength of 250,000
psi at 20°C.

not been tried. They would be tried (and such studies
and alloy development should be supported) if Mn metal
of high quality could be anticipated in the near future.
    Mn Metal, at the right price, would also be used
at lower Mn content levels in steels, where the Mn
compositions would be in the 2-12% range. Mn would
be an excellent substitute for Cr as an alloying
element in low-alloy steels (where corrosion resistance
is not an issue), and also for Ni. In this respect,
one could anticipate large tonnage usage increases
over the current situation.

To develop a Mn technology will require careful
planning, research, and development not only on alloys
and alloy systems, but on extractive metallurgy for
the winning of Co, Cu, Ni, Mn, and perhaps Mo using
Mn nodules. But clearly industry has much to gain
from the added flexibility in raw material supply, in
cost, and in potential for new, interesting, and
useful alloys.

PART III

U.S., THIRD WORLD, AND INDUSTRIAL PERSPECTIVES ON
DEEPSEA MINING

GOVERNMENTAL TREATMENT OF OCEAN MINING INVESTMENT

Arthur Kobler

The U.S. government is very much concerned with the potentially stifling effect of the Informal Composite Negotiating Text (ICNT) on innovation and investment in seabed mining and on the development of seabeds. Spearheaded by the Department of the Interior, the U.S. government is experiencing one of its periodic waves of concern about the adequacy of our current supply of critical raw materials. The Interior Department has formed a nonfuel minerals study which has looked into the domestic and foreign long-term supply situation of critical minerals. The study appears to confirm the conclusion of earlier studies that U.S. reserves of key raw materials are adequate to the nation's demands into the twenty-first century. The major concern is that investments may not be adequate to allow future demands to be met.

Recently the tendency has been for investment funds, especially those used for exploration, to shift away from developing countries, although they may have the greatest potential for the discovery of rich ore bodies. A number of factors have been cited to explain this investment shift. One factor is the basic economic situation involving the weakness in many commodity mineral markets. Another is the political instability in developing countries. But also of great concern are the general investment rules established by most governments, which create a great deal of uncertainty concerning taxes and tenure of contract security. The question of adequacy in investment is important because the deep seabed represents a potentially vast source of raw materials, but it could very well face several of the inhospitable investment conditions that threaten land mineral development. In the negotiations to establish an international mining regime, heated debate arises between factions wanting to create an environment beneficial to private industry and

factions wanting to create a powerful regulatory
mechanism that would operate in a manner similar to
many current governmental land-mining regulatory
bodies.
     In analyzing the exploitation of seabed resources,
including but not limited to the resources in manganese nodules, one encounters several major problems.
First, there seems to be a general tendency to discount projects whose consequences will be realized
in the long term. The seabed resources will probably
not become useful and important to mankind until the
twenty-first century. Current knowledge does not
suggest that seabed minerals will be a necessity for
the world economy for the next 5-10 years. Though a
case might be made for the importance of seabed
minerals in the generations to come, our political
systems (and increasingly, our economic system) are
geared to giving priority to immediate problems.
     Second, in the present U.S. investment approach
to deep ocean mining, one must engage in the unpre-
cedented institutional exercise of creating an international regime for regulating the exploitation of
resources. The outcome of this exercise will
inevitably affect the pattern of resource development
in seabeds. The creation of an international
regulatory regime could also provide a model for the
control of other resources that lie beyond national
jurisdiction, including Antarctica or even outer
space.
     Third, because the seabed is a relatively
unexplored frontier, there is no firm understanding
of the economics of exploiting its resources (present
technology, for example, has yet to be tested). The
best way to find out if commercial exploitation of
the seabed is warranted is to create an environment
in which companies can make rational decisions on the
allocation of investment funds. To the extent that
seabed mining proves competitive with land-based
mining, one would expect the resources to be exploited.
If it proves to be noncompetitive, it is not the

intention of the U.S. government to provide special
subsidies to facilitate the initiation of first-
generation mining projects. This approach is consistent with the U.S. government's overall approach
to other foreign investment.

Fourth, the U.S. government does not intend to
achieve self-sufficiency in mineral production. U.S.
national security needs for strategic materials can
be met by adequate stockpiling, while the needs of
the U.S. domestic economy for assured supplies of raw
materials can be satisfied by encouraging diversified
sources of supply, be they domestic or foreign. With
one or two exceptions (cobalt included), recent
studies have indicated that the United States faces
no major threat of cartels that would seriously dis-
rupt the U.S. minerals market. For the vast majority
of minerals markets, the threat of supply disruption
has been an exaggerated one, stemming from the fears
caused by the 1973-1974 OPEC experience.

It is immaterial to the U.S. government whether
U.S. companies or foreign companies exploit the deep
seabed. The U.S. government's main concern is that
access to the resources be provided on a nondiscriminatory basis with terms and conditions that are
objective and do not distort the process of making
investment decisions. It is generally agreed that an
international regime that allows nondiscriminatory
access to the seabeds on objective terms is preferable
to unilateral action, or action among a few potential
mining states. A set of rules upon which everybody
in the world community at large can agree would provide security of investment and a more certain
investment climate.

For economic reasons as well as for obvious
political and foreign policy reasons, the U.S. government has decided to press the negotiations on an
international regime. Initially, the U.S. government
entered the UNCLOS III negotiations hoping to
establish a very loose arrangement in which the International Seabed Authority would have basically served

as a registrar for state and private claims to mine
the seabed. The Authority would have protected
against overlapping claims and would have enforced
environmental standards. Although a very simple concept, that approach to seabed mining proved unattain-
able in its unadulterated form in the context of the
negotiations.
     Because of the large participation in UNCLOS   III
(over 150 nations), the U.S. government has felt
compelled to make a number of concessions in order
to accommodate the legitimate aspirations of other
participants in the conference. However, in making
these concessions, the U.S. government's basic
objective has remained the same. The United States
hopes to establish a system of assured access for
private miners. Unfortunately the UNCLOS III
negotiations have become entangled in a so-called
North-South dialogue between the developing countries
and the developed countries. As a result, the
negotiations have taken on a larger political context.
     The Group of 77 (now including 115 countries)
has portrayed the seabed negotiations as a model for
their objectives in establishing a new international
economic order. Many of their demands are purely
ideological, such as the establishment of a quasi-
governmental seabed authority which it would control
and which would have complete discretionary authority
to mine the seabeds and to exclude private miners
from the seabeds. Some of the Group of 77's demands,
however, represent legitimate aspirations, and the
United States has tried to accommodate them. Such
aspirations include the desire of the developing
countries to have a significant voice in the Seabed
Authority and their desire to participate directly in
the exploitation of the seabeds rather than leaving
seabed exploitation to the few companies or countries
that happen to have a technological advantage.
     In trying to meet these legitimate concerns and
its own concerns simultaneously, the United States
developed an artful compromise known as the parallel

system. The parallel system involves on the one side
a system of assured access for private and state
miners, and on the other side an International Seabed
Authority with its own operating arm, called the
Enterprise, that will mine the sea independently.
The parallel system seemed an ingenious and eminently
fair compromise, taking into account the different
approaches advocated by the developing countries and
the market-oriented countries. Unfortunately, the
Group of 77 continues to press its political objectives
for a strong central seabed authority with wide discretionary powers to regulate seabed mining. At this
point the negotiations are still far from meeting
U.S. interests in an assured access system for private
miners.

The U.S. government, in coping with the demands
of the developing countries, has encountered a running
controversy over what precisely constitutes an assured
access system. It is open to dispute whether some
U.S. negotiational choices have tipped the scales
away from an assured access system, as opposed to
balancing them. Many companies in the deep ocean
mining industry feel that the former has occurred.

The components of an assured access system
include:

1. A system of exploitation that recognizes the
permanent right of private and state parties to
exploit the seabed.

2. A contractual procedure that is based on an
objective evaluation of the qualifications of applications and that provides nondiscriminatory and
equitable procedures.

3. Revenue-sharing provisions that take into account
the inherent risks and uncertainties of seabed mining
and allow rates of return in keeping with the nature
of the investment.

4. Resource management policies that take into
account environmental concerns and are designed to
allow investment decisions to be made primarily on
the basis of an evaluation of economic criteria.

5. An efficient and fair dispute settlement
mechanism.

6. A decision-making mechanism within the Seabed
Authority that gives major influence to those
countries that have a primary stake in seabed
development.

Even a cursory reading of the current negotiating
text would suggest that these criteria are far from
being satisfied under any reasonable interpretation.
     Two issues with which the U.S. government is
currently grappling and which have a direct bearing
on the issue of defining assured access are the
issues of production control and technology transfer.
The production control issue illustrates the pitfalls
of the UNCLOS III negotiations. Initially the United
States vigorously opposed any kind of production
control on seabed mining because of its inefficiency
and because it would lead to an uneconomical allocation of resources. In 1976, in the midst of intense
negotiations which were hoped to represent the final
stages of the seabed convention, production control
arose as a key issue. The countries in favor of
control were copper producers such as Chile and Peru,
who were concerned with the effects of seabed mining
on the copper market. The most optimistic projections
of seabed mining indicated, however, that copper production from the seabed would represent only about 2
or 3% of total world production by the year 2000.
Although the United States saw this as no threat to
the market, it agreed to production control.
     Despite efforts of the United States to conclude
the negotiations, they were repudiated by the Group
of 77. Cleverly, the Group of 77 managed to pocket

the U.S. concessions on production control. The
Canadians soon thereafter became the leading pro-
ponents of control, not out of benevolence to
developing countries, but out of a desire to protect
their own interests.

In early 1978 the U.S. and Canadian delegations
reached a nonreferendum agreement on production control. The original production control agreement of
1976 allowed seabed mining to fill the completed
projected growth segment of the nickel market for a
period of 20 years. The projected growth segment was
explicitly stipulated to be no less than 6% per year,
which under reasonable projections would not restrict
seabed mining.

After the 1976 agreement fell through, the Group
of 77 and the Canadians decided on a 50-50 split.
The ad referendum agreement followed, providing for
a 60-40 split; 60 for seabed mining and 40 for land-
based mining. It is questionable whether this particular formula will restrict seabed mining or not. It
depends a lot on assumptions made regarding the number
of seabed mining sites that will be out for the next
20 or 25 years and the growth in demand for nickel.

Technology transfer, the second issue with which
the United States is now grappling, has faced an
erosion similar to that of production control. After
the United States proposed the parallel system concept, the LDCs questioned how the Enterprise would
function without capital or technology. Henry
Kissinger provided an answer in 1976 by proposing
that the industrial countries guarantee financing for
the Enterprise's first operation. It was assumed
that the Enterprise, if organized along business
lines, could buy the technology it needed by entering
into a joint venture agreement or by hiring an
engineering company to develop the technology from
scratch. The United States was prepared to agree to
conditions that would encourage joint ventures with
private miners or the Enterprise on nondiscriminatory
terms and would promote training programs to help

developing countries acquire necessary mining skills.
The Group of 77 then shifted their focus from
financing to technology, claiming that the companies
might not sell the technology to the Enterprise, As
a result, the LDCs began demanding provisions
requiring the transfer of technology to the Enterprise.
(The Group of 77 has been pursuing similar ideological
objectives with regard to mandatory technology transfer in other U.N. forums.) The United States feels
that obligatory technology transfer is unnecessary.
The Enterprise will have substantial capital resources
provided or guaranteed. The United States feels that
if it is organized along business lines, there is
little question that the Enterprise will have the
capacity to enter a joint agreement or to develop the
technology on its own. At present the United States
is seeking a solution that goes beyong simple commercial purchase of the mining technology but not as far
as mandating technology transfer.

In conclusion, let me simply reiterate that the
major objective of the United States in the seabed
negotiations is to resolve the various disputes,
such as technology transfer and production control,
in such a way as to guarantee assured access and
create a hospitable climate for investment in the
seabeds.

U.S. DEEPSEA MINING POLICY: THE PATTERN AND THE PROSPECTS

Richard G. Darman

INTRODUCTION

This paper examines the pattern of U.S. policy development with respect to deepsea mining. On the basis of this pattern, it projects the likely further evolution of U.S. policy. And it frames an analysis of the consequences of that policy from the perspective of U.S. interests.

From the author's personal viewpoint, the observable pattern and prospects seem regrettable. Accordingly, the analysis of U.S. policy development has a relatively negative cast. Any such analysis, however, is influenced by one's evaluation of "U.S. interests"; and there is obviously much room for debate about these. It is perhaps important, therefore, to note that although the analysis is influenced by the author's personal evaluation of U.S. interests--an evaluation that gives relatively heavy weight to the precedential context in which the deepsea mining regime is being developed, and one that tends to prefer market-oriented economic systems and decentralized, relatively democratic, political systems-- the prognosis here is neither dire no bleak. It is simply (and, of course, arguably) less favorable than it might have been had an alternative line of policy development been pursued.[1]

Whatever one's personal perspective as to the proper weighting of U.S. interests, it is useful to be clear about several major points of objective perspective that help describe the environment in which U.S. policy has developed, and in relation to which it must be assessed. Six such points of perspective might be highlighted as follows.

First, the deep seabed area amounts to half the earth's surface, a vast territory by any conventional reckoning. The fact that it happens to be covered with water may mislead the unimaginative toward an underestimation of its potential value, just as snow

and ice have done in the cases of Alaska and
Antarctica, just as heat and apparent barrenness have
done in the case of Mexico, just as remoteness and
disparateness have done in the case of Micronesia, or
just as water itself has done with regard to many
emergent ocean uses. But on conventional territorial
grounds alone, without adjustment for possible fail-
ures of imagination, the establishment of a governmental regime for such a vast area would have to be
deemed highly significant.[2]

Second, the value of the area and the importance
of the associated regime cannot properly be calculated on the basis merely of some discounted present
value estimate of the economically interesting
minerals. This is true partly because other ocean
uses are also at stake; partly because the regime
may have important political consequences well beyond
deepsea mining and well beyond the oceans; and,
perhaps most importantly, because the fundamental
values at stake involve future generations--a fact
that complicates considerably both the technical and
the ethical analysis of current policy making.

Third, in the course of the U.N. negotiations
concerning a deep seabed regime, the possible political significance of this regime--beyond ocean mining--
has risen considerably. The notion of the "common
heritage of mankind" has gained appeal with respect
to the "global commons" generally and also with
regard to outer space. Global political consciousness
among the many new countries of the world community
has been dramatically awakened. The "North-South
dialogue" and the quest for a "New International
Economic Order" have become more intense. And in
general, for a variety of reasons (ranging from a
perceived need to rationalize the proliferation of
less-than-satisfactory international organizations to
a perceived need to limit unilateral actions of
states), the interest in global institution-building
and reform has increased.

Fourth, in this context it is particularly
important to note that the deepsea mining regime now
being negotiated has a potential "reality" to it,
along several relevant dimensions, that distinguishes
it from many subjects treated by U.N. conferences.
The proposed regime will take form as a largely free-
standing and operational politico-economic system--
with an ongoing representative governing body,
independent regulatory powers, an associated judicial
system, independent revenue-raising capability,
independent revenue-dispensing authority, and a new
globally chartered commercial Enterprise. For those
interested in global institution-building and reform,
it would be difficult to find a more exciting focus,
among those now realistically available, than that
provided by the opportunity to design and develop a
deepsea mining regime. It is important also to note
that, for this and other reasons, the negotiation of
a deepsea mining regime is highly visible in the
"North-South" context, especially among the "Group
of 77" developing countries.[3]

Fifth, whatever the value and importance of the
deep seabed regime, it has consistently been viewed
by the United States as only one among several
important oceans interests thought to be at stake in
the law of the sea negotiations--the most important
of which, according to the guiding conventional
wisdom, is the preservation of certain traditional
high seas freedoms of commercial and military navigation. The relative importance of these freedoms is
arguable; the relative importance of their universal
protection by treaty is arguable; and the extent to
which the proposed treaty language is consistent with
their protection is arguable. But the fact that
treaty protection of these freedoms by the Executive
branch has been a principal objective of U.S. participation in comprehensive law of the sea negotiations
is not arguable.[4]

Sixth, as a final point of introductory
perspective, it is useful to note that if one were

interested only in the requirements for a responsible
legal framework within which deepsea mining could
healthily develop, there would be no inherent need
for the type of international governmental and
economic system being negotiated within the U.N.
framework. The current legal status of the seabed
area does not require such a system: traditional
high seas freedoms still obtain (notwithstanding a
nonbinding U.N. resolution to the contrary), and
necessary legal protections can be provided through
sponsoring-state regulation under the principle of
"nationality" jurisdiction (i.e., based on a state's
authority to regulate its citizens and corporations
wherever they may be).[5] Security of tenure for
miners and investors does not require such a system:
for this purpose, reciprocal agreements among mining
countries would suffice. Environmental protection
does not require such a system: necessary environmental regulations could be negotiated, and periodi-
cally updated, by the states operating under reciprocal agreements. Efficiency does not require such a
system: the system being negotiated will reduce
likely efficiency. And even equity does not require
such a system: revenue sharing and participation by
developing countries could be assured by inclusion of
relatively simple tax and transfer and subsidy
measures in a framework of domestic legislation and
reciprocal agreements.[6]
     Indeed, as suggested below, the type of regime
now being negotiated cannot be justified from the
perspective of U.S. (or even global) interest in
optimal seabed resource development. If the regime
is to be justified from a U.S. perspective, it must
be with reference to interests that have little
inherent, direct connection whatsoever with deepsea
mining.

THE PATTERN OF U.S. POLICY EVOLUTION
For the past decade--that is, for the history of the
preparations for and negotiations of a deepsea mining

regime--U.S. deepsea mining policy has been consistently oriented in one general direction. It has
sought to develop universal agreement on a mining
regime within the broader context of comprehensive
law of the sea treaty negotiations conducted under
the auspices of the United Nations. The United
States might have done otherwise. It might have pro-
ceeded through less-than-universal agreement, or a
less-than-comprehensive treaty, or non-U.N. auspices.
But it did not--a point to be discussed further
below.

Within this consistently affirmed general framework for the development of a deepsea mining regime,
U.S. policy has shown relatively little substantive
consistency--other than a clear pattern of concession
to the policy preferences of the Group of 77, concession that characteristically has been unilateral.

As a theoretical matter, one might conceive of
several alternative seabed regimes--ranging from the
more decentralized to the more centralized and from
those more naturally consistent with U.S. ideology to
those more consistent with the composite ideology of
the Group of 77. This range moves from a simple,
decentralized, minimally conditioned licensing
regime; to a more centralized variant of the same; to
a so-called clean parallel system; to a so-called
balanced development system; to a so-called mixed
system (adding joint ventures and more central negotiating discretion to the "parallel" and "balanced"
concepts); to a unitary joint venture system with
clear, continuing rights of access for state-sponsored
parties; and on to a unitary system without such
rights of access. And just as the mind can move
rather easily from the relatively free or decentralized end of this range to the other, U.S. policy has
moved along this continuum--although it is not yet
fully at the last centrally regulated stop.[7]

Key points in the movement of U.S. policy might
be highlighted as follows.[8] At the outset--indeed,
in 1970, before the formal negotiations even

commenced--the United States skipped over the
decentralized licensing alternative and unilaterally
proposed a draft convention with a centralized (but
still minimally conditioned) licensing regime, replete
with an International Seabed Resource Authority and
with associated Assembly and Council machinery, and
with generous provision for revenue sharing of all
mineral resources beyond the 200 meter isobath.

At the same time, the United States affirmed the
"common heritage of mankind" principle, giving
impetus to the U.N. resolution on the subject--a
resolution that was passed, with U.S. support,
shortly thereafter. (The "common heritage" notion
had, of course, been a subject of considerable U.N.
interest since Maltese Ambassador Arvid Pardo's 1967
address. And the United States had given prior
rhetorical support to a variant of the notion in a
speech by President Lyndon Johnson. In 1970, however,
the notion was agreed to as a governing principle--
leaving its operational meaning to be determined.)

Insofar as the United States might have received
something in return for these initiatives, the
"return" might be said to be the comprehensive law
of the sea treaty negotiating framework. But as the
negotiations (and evolving customary law) have been
played out, there is at least reason to question
whether the United States may have established at
the outset a framework with an inexorable logic to
it--a logic that had to be adverse, at a minimum, to
U.S. interests in shaping a seabed regime. At the
very starting point, the United States ceded a fundamental institutional issue (that there should be a
significant new Authority) and a fundamental issue of
principle ("the common heritage of mankind")--in
return for which it got a negotiating framework that
it hoped would allow it to get certain economic zone
freedoms in trade. In fact that framework merely
assured that the United States would have to trade
the operational provisions of a centralized seabed
regime in its attempt to secure economic zone freedoms

that were subject to increasing erosion and to
sustained threat of further erosion.

When the formal seabed negotiations commenced in
1974 and did, in fact, yield a highly unsatisfactory
negotiating text, the United States decided to offer
further concessions in an effort to gain improvements
in a revised negotiating text. The ensuing process
of negotiation was to move further and further away
from the minimally conditioned licensing regime that
the United States had originally envisioned. In 1975
the United States accepted the idea of a "parallel
system," which on the basis of private negotiations,
it thought would be a widely acceptable compromise.
(It was in fact promptly rejected, upon its general
exposure, by the Group of 77.) And the United States
began to introduce the principle of "balance" in the
proposed parallel system by volunteering that miners
should provide prospected minesites, free, to an
internationally chartered competitor, "the Enterprise."

In the same context, in order to secure agreement
on the parallel system, the United States began to
negotiate a formal production control system (to protect land-based producers). And in 1976 the United
States volunteered (in a major speech by the Secretary
of State) a specific production control formula. The
formula was intended to be a "nonbite" production
control (and, if adopted, would likely have been
such). But the issue of principle was formally ceded;
and it was only a matter of time (until 1978, as it
happened) before the dynamics of negotiation caught
up, and the United States agreed to a stringent and
arbitrary production control formula.

In a continued vigorous quest for agreement (still
in 1976), the United States moved further toward what
was later to be termed a "balanced development system."
In a highly visible "breakthrough" effort by its
Secretary of State, the United States volunteered to
assure financing for the start-up of the proposed
"Enterprise" (which might require as much as three-
quarters of a billion dollars of underwriting) and

also the necessary transfer of technology. Both
offers were general when made, but the negotiating
dynamics soon made them quite particular. (in 1977
the United States attempted to regain some ground by
making the financing offer a shared debt-underwriting
offer and by arguing that this would be sufficient to
assure technology transfer. In the same context, the
United States unequivocally opposed any mandatory
system of technology transfer. But the negotiating
text produced by the Conference included both the
U.S. financing proposal and a mandatory system of
technology transfer, which was accepted by the U.S.
negotiators in 1978.)

   As part of the 1976 "breakthrough" effort, the
United States also proposed that the parallel system
be subject to review in 20-25 years; and in 1977 it
reaffirmed this proposal, subject to acceptance of
the parallel system as a compromise, and provided
that the terms of review not be prejudiced in favor
of any alternative system. By 1978, however, U.S.
negotiators--who had long since moved beyond the
parallel system--moved to accept a review system
prejudiced in favor of those with biases toward a
unitary system or biases against seabed development.
(There is a particularly neat irony here. The United
States initially proposed review along with financing
of the Enterprise, technology transfer, and production control in order to achieve what was a compromise
from its perspective, that is, the parallel system
rather than its preferred licensing system. But its
concessions, even when "sweetened," did not gain the
compromise. All they did was assure that the Group
of 77, or some such successor organization, would be
in the best possible position at the time of review.
And nonetheless, the United States saw fit to agree
to concede important elements of its review position
to move it further in the direction advocated by the
Group of 77.)

   In 1977 the United States began to show a
willingness to negotiate a "mixed" regime, allowing

incentives for joint ventures with the Enterprise and
allowing the Authority to enjoy certain appearances
of broader discretion--provided that these were to be
clearly and satisfactorily limited by other provisions, specifications, and references in the text and
its annexes. By 1978 it was taken for granted that
the regime would be no less "unitary" than a "mixed"
regime. But the negotiating text and annexes had
failed to be satisfactorily limited; and the United
States began to reconsider whether it should really
seek such extensive limitations.[9]

In addition to this pattern of concessions moving
rather directly along the continuum of possible
regimes, it is important to note that the United
States has explicitly and implicitly made many other
concessions that do not neatly fit this pattern--but
fit only the emerging pattern of a relentless pursuit
of a comprehensive treaty. For example, the United
States has accepted language creating a "supreme
Assembly" on the basis of one-nation/one-vote,
although it has yet to agree to the operational meaning of the contemplated supremacy.[10] Or as another,
somewhat more obscure, example: the United States
has willingly (although some claim accidentally)
participated in the drafting of language that is now
in the text and that could oblige it to accept an
effective and indefinite moratorium on the development of all seabed mineral resources other than
manganese nodules.[11]

In return for all these concessions, the United
States has received (at this writing): textual
incorporation of its concessions (or even more concessional variations thereof); no substantive concessions from the Group of 77;[12] an outright rejec-
tion of the parallel system; a series of essentially
unchanged demands from the Group of 77; and expression
by the Group of 77 of a continued "willingness to
negotiate." For the latter, the United States has
seemed to be most appreciative.

It is important to note finally, in completing
this sketch, what the United States has done with
respect to domestic legislation concerning deepsea
mining. Such legislation has been intended: to be
consistent with (and superceded by) any U.N. treaty
ultimately agreed and ratified; to be consistent with
existing international law; and to be capable of
providing a framework that would allow seabed investment and technology development to move forward,
while assuring necessary protection of the ocean
environment. But such legislation has been viewed
as offensive by developing countries, who consider it
inconsistent with both the "common heritage" principle and the U.N. seabed moratorium resolution.
(See note 5.) The executive branch has therefore
been opposed to domestic legislation--until late
1977, at which point the "fundamentally unacceptable"
character of the negotiations necessitated a domestic
shift in policy. In 1978 legislation overwhelmingly
passed the House, but was stalled in the Senate at
the end of the session. Current prospects are
uncertain--although it seems likely that legislation
will continue its slow move forward toward enactment.
Whether enactment would affect the Conference
adversely or favorably is a debatable matter. For
the moment, the simple fact remains that the United
States has not moved to establish a framework--even
an interim framework--of state practice under domestic
legislation. It has explicitly and implicitly given
greater weight to its interest in the conclusion of
a comprehensive treaty.

This general pattern of U.S. concessional
behavior does not lend itself to rational explanation
if one's frame of reference involves only seabed
mining. Within that narrow framework, there is not
only a lack of objective requirement on the merits
for the type of regime the United States is participating in developing (as noted above). But there is
also a lack of negotiating requirements for the type
of regime the United States is developing, for with

respect to seabed mining alone, the United States
(and its British, French, German, and Japanese allies)
have, throughout the negotiations, had virtually all
the essential leverage. (The leverage may not have
been, and is not now, sufficient to develop a satisfactory regime within the U.N. negotiating framework.
But, as discussed further below, the United States
did not have to choose to enter, or to remain within,
the U.N. negotiating framework.)

The general pattern of continued U.S. participation in the U.N. Conference and continuous U.S.
concessions (that is, above and beyond the ordinary
amount of tactical blunder) can only be understood by
reference to interests beyond seabed mining, and
their seemingly relentless pursuit through a strategy
of comprehensive law of the sea negotiations.

LIKELY NEXT STEPS IN THE SEABED NEGOTIATIONS
There are two major seabed issue areas that have not
yet (as of early 1979) been the subject of serious
and general negotiating attention within the U.N.
Conference. This inattention has not been accidental,
since both issue areas have been thought to be
potential "conference-breakers"; and the Conference
leadership has postponed their serious consideration
while awaiting progress toward consensus on other
issues--hoping to be able to take advantage of
"momentum" and "closeness" to agreement in order to
avoid a stalemate or breakdown. These issue areas
are both important and inescapable, and they are
bound to gain more serious attention in the near term.

One of these issue areas involves the powers,
composition, and voting of the Assembly and Council.
The United States has consistently maintained that
issues in this area must be resolved in a manner that
(1) will assure the expeditious approval of "quali-
fied" applicants (as understood by the United States),
and (2) will also provide the United States (or a
combination of the United States and a very confi-
dently predictable set of allies) with the essential

voting power to prevent adverse rulemaking and
regulating by the proposed Authority. From the
perspective of the Group of 77, these conditions are
ideologically offensive. In their view, the first
condition unduly denies appropriate discretion to the
international Authority; and the second condition--
which has been understood as one form or another of
U.S. veto system--is inconsistent with the fundamental
political principles of the intended "New International Economic Order" and smacks of the much-lamented
U.N. Security Council veto system. In the U.S. view,
by contrast, these have not been treated as ideological issues (although lately they have been increasingly
understood to have precedential significance), but,
rather, they have been treated as issues of essentially pragmatic concern.

In a contest between a fundamental ideological
concern and a concern perceived as pragmatic and
part of a larger pragmatic framework, it would seem
reasonable to predict further "pragmatic" concession--
in this case resulting in a system that gives considerable discretion to the international Authority,
supreme policy making responsibility to a one-nation/
one-vote Assembly, and clear veto power only to
developing countries.

The second difficult issue area still to be
treated seriously in general negotiation involves
provision for quotas and the equitable distribution
of contracts among classes of applicants and states.
The United States has been opposed to any such provisions that might have further restrictive effect
with respect to U.S.-sponsored applicants, arguing
that, under the proposed system, at least half of all
minesites are to be reserved for the Enterprise and
developing countries and that global interests would
best be served by the application of relatively free
market principles to the remaining minesites.

This U.S. position has been resisted by other
developed countries, who are concerned that it might
gain too many minesites at their expense. When

debated in open forum, preference for quotas and
"equitable" contract distribution systems can
reasonably be expected to gain relatively widespread
developing-country support--in part on grounds of
appropriate discretion for the Authority, and in part
on grounds of "equity" in the division of the "common
heritage" (revenue sharing, "banking," financing the
Enterprise, etc., not being considered sufficiently
equitable in this respect). But because this issue
area is not of fundamental ideological concern to the
developing countries, there has been, until recently,
at least the theoretical possibility of negotiating
an acceptable resolution of this issue among key
developed-country mining states. Lately, however,
the issues have been seriously complicated by the
U.S. concession of a stringent production ceiling
formula. And having made that concession, it would
appear that the United States may be driven inexorably
to some further restrictive concession under the
heading of quotas or equitable contract distribution.

In addition to these two important issue areas
that remain to be negotiated generally and seriously,
there are of course others in which ongoing negotiations have not yet identified the outlines of an
implicit consensus, and in which the United States
may be called upon to make further concessions. One
of the most important of these involves the financial
terms of contracts, in which, as some have observed,
the Conference seems to be searching for "the minimum
ratifiable return on investment." On these and other
such important matters of detail, the patience of the
Conference (and not least, the United States) is
being tested. [13]

One approach to the problem of resolving remaining
issues of detail--an approach that seems to have
gained some favor recently among U.S. negotiators--is
to invent one or another procedure for temporarily
freezing what is agreed and shifting issues of detail
to some successor negotiating forum, typically conceived as a more technical interim body. This

recently emergent notion has presumed that only when
details have been ironed out would the entire package
be subject to ratification procedures. Upon analysis,
variations of this approach turn out to seem either
trivial in their significance, or highly unlikely to
gain Conference approval, or likely only to consoli-
date U.S. losses while establishing a framework
through which to lose even more. For this reason, it
is here assumed that this approach will not be pursued
by the United States as a means to alter substantially
and favorably the basic deepsea mining texts--while it
nonetheless may be pursued with respect to some
second- and third-order issues of detail.[14]

     In sum, given the pattern noted above and many
other considerations, it seems reasonable to expect
that any deepsea mining regime that emerges from the
present U.N. Conference will have most or all of the
characteristics that the United States deemed "fundamentally unacceptable" following the issuance of the
Informal Composite Negotiating Text (ICNT): but it
also seems reasonable to expect that the U.S. negotiators might nonetheless find such a regime acceptable, when viewed within what the United States has
taken to be the broader law of the sea context.

A PATH NOT TAKEN

The foregoing exposition might correctly be read to
suggest an inexorability to the course the United
States set for itself. It is important, however--at
least analytically--to note that one might define a
plausible alternative. This course would be more
decentralized, less comprehensive, and less universal
(at least for the next 20-25 years) than the course
the United States has actually followed, and from
which it seems unable to get itself to depart.

     In brief, the essential elements of this alter-
native course would be comprised of the following:

1.   prompt enactment of domestic legislation along
the lines of HR 3350 as it passed the House of

Representatives in 1978, and prompt issuance of
permits under such an act (in part to encourage sound
seabed development, and in part to establish a
favorable pattern of state practice);

2.   inclusion in the legislation not only of provisions for ultimate revenue sharing with the poorest
of nations, but also of provisions to encourage
developing-country participation in deepsea mining
consortia--as might be done, for example, by diverting
revenue-sharing payments in the near term to subsidize
interest on developing-country debt obtained for the
purchase of equity interests in joint ventures with
developed-country miners (in part for reasons of
equity, in part for reasons of practical merit, and
in part for reasons of politics and appearance);

3.   vigorous encouragement of reciprocating agreements and harmonious domestic legislation among
developed-country mining states (to assure security
of tenure and responsible state practice); and

4.   vigorous encouragement of selected developing-
country participation in the joint arrangements
contemplated by point 2 (to assure that the provisions are not just for appearance' sake).

     Pursuit of this course obviously would require a
greater preference (at this stage of global development) for decentralized institutions, or a greater
relative weighting of U.S. seabed interests, or a
lower assessment of the probability of success within
the Conference framework than has been characteristic
among U.S. negotiators.
     Pursuit of such a course might also be based on
the reasonable judgment that the current international
climate is not propitious for the negotiation of
universal norms by all the nations of the world--that
the times are too close to the end of the colonial
era, to the exposure of abuses of power by multinational

corporations, to Vietnam, and to other reasons for
associating developed countries with imperialism and
market-oriented economies with inequity; that the
excessive reactive positions of developing countries
are not now consistent with the best interests of the
global community; that the reduction-to-lowest-common-
denominator problem is extreme; that there is no com-
pelling reason for the United States to accept such
reductionism (indeed, that the United States may have
a responsibility not to); and that with time and more
responsible state experience, the international
climate may become more favorable for the negotiation
of universal norms than it is today. Quite obviously,
these judgments have also not characteristically been
held by U.S. negotiators.

In addition to these differences of preference and
judgment that help explain why the United States has
not followed the alternative course outlined above,
there are two other elements of possible explanation
that are particularly noteworthy.

One of these is a matter of legal analysis.
Although it is not the official U.S. legal position,
there is a significant undercurrent of opinion that
the United States might be at risk legally if, absent
a treaty, it exercises its high seas freedom to mine
the seabed within a framework of domestic legislation
and reciprocal agreements. This point of view is
especially noteworthy because the legal analysis would
seem to presume that the United States should or would
accept as binding a legal opinion based primarily on
hitherto nonbinding U.N. resolutions. And this, of
course, would be a radical shift of U.S. policy with
regard to international law and international
institutions--a policy shift that would far transcend
the law of the sea. (It would be a major step toward
granting effective legislating power to the U.N.
General Assembly.)

The second additional explanation for the failure
of the United States to follow the alternative course
outlined above is the opposite of radical. It is,

simply, inertia. It bears noting because it is very considerable--and, as a matter of realism, it should be presumed to reduce dramatically the probability of any objective analysis having the power to shift significantly the current U.S. course.

Indeed, the only events likely to shift the U.S. course are rather extreme in the context of the law of the sea. These are (1) Conference breakdown (or the unequivocal prospect thereof), notwithstanding the pattern of U.S. concessions, or (2) nonratification by the U.S. Senate.

Both of these, assessed realistically, must be viewed as low-probability events. The former is improbable in part because it is in the developing countries' perceived interest to keep the Conference going, and in part because the U.S. pattern of concessions reinforces that perception. The latter is improbable,[15] notwithstanding the pattern of U.S. concessions, for a variety of reasons. Important among these are the combined political power of a President, Secretary of State, Secretary of Defense, and Chairman of the Joint Chiefs of Staff (who would support the treaty on grounds not related to seabeds); the romantic appeal of international institution-building and "sharing the wealth," independent of detailed regard for the means of so doing; the inherent political weakness of an infant industry; and the general lack, in the United States, of a significant political or institutional constituency effectively concerned with events that have uncertain tangible connection with the near term, and yet have what may be fundamental connection with the more distant future.

A FRAMEWORK FOR THE EVALUATION OF LIKELY IMPACTS (FROM THE PERSPECTIVE OF U.S. INTERESTS

As suggested above, a deepsea mining regime can be outlined quite clearly by describing and extending the pattern of U.S. concessions within the U.N. Conference negotiations. In the author's view, the

regime thus described should be taken as by far the
most probable ultimate outcome. It should, therefore,
be the principal focus of evaluative concern.

The question immediately arises, however, as to
what regime(s) to use as comparative reference(s).
There is, of course, a near-limitless range of
theoretical possibilities. But, for better or worse,
the evolution of negotiating realities has removed
most of these and thus dratically simplified the task
of analysis. (In the author's view, there is only one
basic plausible alternative to use for comparative
analysis: a form of seabed "minitreaty" comprised
essentially of mining states' domestic legislation
harmonized through reciprocal agreements. (The two
basic variants of this alternative are the one
described above as "a path not taken" and the identical regime without the provision for developing-
country participation in joint ventures.)

It should be noted that there is a realistic
possibility of a temporary (formal or, more likely,
informal) seabed mining moratorium, pending further
negotiations. But any such moratorium would not be
likely to slow seabed development by more than a few
years. Since the key U.S. interests relate to the
more distant future, and since the movement of near-
term events plus or minus a few years does not (of
itself) fundamentally affect the analysis of long-
term interests, near-term moratorium variants can be
put aside for purposes of simplified analysis.

Taking the projected, highly concessional,
centralized regime as the basic focus, and taking the
decentralized "path not taken" as the implicit
comparative referent, it is possible to indicate the
basic direction of evaluation--"favorable" versus
"unfavorable"--with respect to U.S. interests, as
follows.

"Favorable" effects of the projected seabed
regime would include:

1. The negotiation of a comprehensive law of the
sea treaty that would protect or promote certain non-
seabed interests that the United States attaches to
the law of the sea. The degree to which the likely
comprehensive treaty would, in fact, protect or promote such interests, as well as the relative importance of such interests, are arguable matters.

Among other interests ordinarily thought to be at
stake in the comprehensive treaty negotiations (with
the author's view noted parenthetically) are fishing
interests (which, during the extended U.N. negotiations, have been largely overtaken by the development
of customary international law); interests in the
development of compulsory dispute settlement procedures (which, in the current negotiating texts, are
severely flawed and largely illusory); interests in
environmental protection (for which the contemplated
treaty provides what is essentially a framework for
future standard-setting, enforcement, and negotiation--
not obviously likely to be very different in effect
than the pattern to be expected in the absence of a
treaty); and interests in the preservation of certain
traditional high seas freedoms of military and
commercial navigation (some of which appear to have
been overtaken by recent history, others of which do
not necessarily require comprehensive treaty protection, and still others of which seem unlikely to be
protected by treaty in any case).[16]

2. A widely accepted, and in some respects a
somewhat more stable, legal framework for deepsea
mining. (While a widely ratified treaty would
undoubtedly have this advantage, it is important to
note that there are respects in which the framework
may be less than satisfactorily stable. These
involve remaining uncertainties as to the extent to
which there will be satisfactory protections against
abuses of power by the new Authority--abuses that
could either destabilize contractual relations or
destabilize the political support for the Authority.

And they involve the inherent uncertainty of the
required review process.)

Unfavorable effects of the projected seabeds
regime would include:

1. In aggregate, a reduced rate of investment in
and development of deep seabed resources (a) with
respect to minerals in nodules and (b) with respect
to other resources. The reduced rate with respect to
both (a) and (b) can be expected because of increased
uncertainties, increased requirements for negotiating,
increased production controls (although the aggregate
level of allowable production may not necessarily
differ significantly from market-regulated levels),
and decreased allowable financial return to investors.
There is also a special problem with respect to seabed
resources other than those to be found in nodules--in
that they may be subject to a legally binding moratorium, pending the future negotiation and ratification of appropriate rules and regulations. (See note
11.)
2. For U.S. companies in particular, a reduced
share of aggregate investment in and earnings from
seabed investment. Under the more decentralized
regime, U.S. companies would, of course (as they
already have to some extent), seek foreign partners
to reduce their political risk, thereby reducing
their share of aggregate investment. But under the
centralized regime, U.S. companies would also be
rendered ineligible for direct investment in at least
half the minesites (although they might nonetheless
operate on some of these sites, in the near term, as
subcontractors); and they would be subject to whatever quote/equitable-contract-distribution system is
adopted, further reducing their opportunities for
investment.
3. Because of points 1 and 2, and because the
projected regime would give wide discretion to the
new Authority, reduced surety of U.S. access to the

mineral resources of the seabed. This is a significant
problem only for the longer term, and only assuming
that the United States could not then adequately and
economically stockpile necessary minerals purchased
directly or indirectly from the Authority and other
producers. The most obvious potential problem
identifiable now involves manganese, which is, using
present technology, essential for the manufacture of
steel, and secure sources of which, according to
some estimates, are in danger of being exhausted
within 25-30 years.[17]

4. Reduced incentives for technological development. This is partly a function of the same variables
as cause 1, and partly also a function of the mandatory technology transfer regime. It should be
noted, however, that the effect of the latter alone
might be only to change the nature of technology
developers from firms dominated by mining interests
to firms governed by interests in technology transfer.
Whether incentives for such firms would be reduced or
not would depend on the terms negotiated with the
Authority. But the very uncertainty associated with
such negotiations would seem likely to reduce incentives for technological development to some extent.

5. Increased prices of the minerals involved.
This is partly a function of causes 1, 3, and 4. The
problem is compounded by the fact that for some of
the minerals involved, production controlled by the
proposed Authority would (in time) represent a major
proportion of global production--and the Authority
would be in a position to operate as a cartel manager.

6. A reduction in the resources available for
global redistribution of wealth. This is principally
a function of the reduction in aggregate investment
and in the incentives for technological development.
It should be noted, however, that resources "available"
for redistribution are not necessarily subject to
redistribution, and that insofar as the proposed
Authority might develop significant power, it might
be able to extract monopoly rents that would more

than offset the presumed losses. It should also be
noted that under any alternative now realistically
available, including the one being developed by the
U.N. Conference, the near-term redistributive effects
are likely to be trivial.[18]

   7. A reduction in high seas freedoms in the deep
ocean area. The proposed Authority would not only
regulate seabed mineral resource development, with
broad powers for related matters of environmental
protection, but could very reasonably be expected to
expand its scope of jurisdiction in time to other
ocean uses (notwithstanding would-be limiting language
in the present text). Indeed, this expansion would
be the natural line of development (and the burden of
argument to the contrary should rest with those who
argue that it would be otherwise).

   8. Adverse precedential effect with respect to
the development of global political and economic
institutions generally. In virtually every major
respect--the form of economic controls, the degree
of centralization, the lack of protection of minority
interests, the distance from fair representative
democracy, the inadequacy of judicial protection,
etc.--the projected regime is fundamentally anti-
thetical to American values and, if generalized, to
American interests. Whether or not acceptance of
the adverse deepsea mining precedents would, in fact,
increase the chances of their being generalized--and
if so, by how much--is debatable. But it is clear
that for all the reasons noted above, the global
community is increasingly interested in global
institution-building and reform; and the law of the
sea negotiations are particularly visible in this
context.[19]

   It should be noted that, although the list of
unfavorable effects is long, the author does not
reach the extreme conclusion that some advocates
have: namely, that under the projected mining
regime there would be "no mining." A reasonable

presumption, rather, would seem to be that once the
Authority's likely reality were recognized (however
unpleasant it might be to contemplate in the abstract),
a set of would-be miners would find their way to its
door; that with legal control of the seabed and with
financing, the Authority would get itself "in business"'
that once in business, it would seek to remain so.

But even though there would be mining, there
would, as suggested, be problematic effects on price,
access, technology development, global efficiency,
equity, and the shape of future global institutions
generally. These are offset, to some arguable
degree, by the posited benefits. Some of these costs
and benefits might be quantified. But unfortunately,
the ones that might be quantified most easily are, on
the whole, trivial. The least trivial and most
difficult to assess are the non-nodule interests at
stake--including both those directly at stake in the
comprehensive treaty negotiations (such as certain
freedoms of military navigation) and those that might
be at stake through the indirect and precedential
effects of the projected seabed regime (such as the
character of future global political and economic
institutions).

For the past several years, U.S. policy has
resisted any systematic effort to assess interests,
costs, and benefits. U.S. policy has seemed to
trust, in part, that the benefits associated with
navigational interests and with agreement per se
might outweigh all likely costs; and U.S. policymakers
have wished, perhaps, to postpone the need to address
explicitly the difficult domestic issues of cross-
interest "trades." It is clear now, however, that
within the U.N. negotiating context, trades of one
set of interests for another are inescapable. But
it would appear that the explicit U.S. assessment of
these trades will not be made until the process of
ratification is engaged.

In some respects, as suggested above, this is
regrettable. The ratification process is hardly a

sensitive and flexible mechanism for policy development. Indeed, flexibility is so narrowed by the
stage of ratification that the probability of ratification (which may be moderately high in any case) is
increased. In the end, however, this is not objectionable. For the key interests and assessments
involve such difficult issues of value and judgment
that the best calculus for their assessment may well
be a rather clumsy, binary, democratic political one.

NOTES
1. For the author's assessment of U.S. interests in
law of the sea, see Richard G. Darman, "The Law of
the Sea: Rethinking U.S. Interests," Foreign Affairs,
vol. 56, no. 2 (January 1978), and Richard G. Darman,
"Statement: Precedential Implications of a Deep
Seabed Mining Regime," in Hearings, U.S. House of
Representatives, Subcommittee on Oceanography, serial
no. 95-44 (Washington: GPO, 1978). For a related
assessment of the implications for policy choice, see
Richard G. Darman, "Choices in Law of the Sea Negotiations: An Analytic Framework and Personal Assessment," in The Oceans and U.S. Foreign Policy
(Charlottesville: University of Virginia Center for
Oceans Law and Policy, 1978).

2. The United States--influenced importantly by its
concern for the preservation of certain traditional
high seas freedoms of navigation and overflight--has
tended to resist any such territorial conception or
formulation of interests in the deep seabed area.
That such territorial interests are at stake, however,
is indisputable. As the deep seabed area is developed,
various forms of de facto control (even if only
temporary) must rest with some combination of state-
sponsored entities and a variant of the proposed
international Authority. The United States has, until
recently, sought to protect its seabeds interests
through a system that would "guarantee" or "reasonably
assure" its access to the deep seabed, while insisting

that no system should interfere with traditional high
seas freedoms in the superadjacent waters. It is
noteworthy that the USSR has shown a consistent,
traditional sensitivity to the possible geopolitical
implications of seabed territorial control--reflected,
on the one hand, in the Soviets' positive interest in
quota and antimonopoly provisons (to limit potential
U.S. control) and, on the other hand, in the Soviets'
interest in assuring itself quasi-veto power in the
Authority (to limit potential control by any group of
states through the proposed Authority).

3. The relative invisibility of the law of the sea
within the United States generally, and even within
the U.S. foreign policy community, should not be
mistaken for an appropriate representation of the
visibility that the law of the sea commands among the
majority of states. Law of the sea ranks particularly
highly on the policy agenda of the "Group of 77,"
which now numbers more than 115 states. These states
are, of course, remarkably diverse along most
dimensions--even including per capita income (ranging
from Saudi Arabia and Kuwait, whose per capita incomes
are among the highest in the world, to states that
are clearly the poorest by this measure). But
although their domestic political systems also vary
widely, they share an expressed ideological commit-
ment to the New International Economic Order, bound
by a common sense that their development has been
undesirably affected by injustices in the current
distribution of power and wealth among states and by
injustices in the global politico-economic system
that has produced and supported this distribution.
The Group of 77 now routinely meets to develop
unified policies and negotiating positions for major
global negotiations. And notwithstanding the considerable differences and splits within the group,
"solidarity" has been relatively well maintained,
both within and across particular negotiations.
(This fact may result in part from "the North's"

tactical decision to treat both the Group of 77 and
itself as if each were "solid," which of course is
not the case for either. The Group of 77's interest
in the law of the sea is partly to be explained by
the particular interests of some of its members. But
the subject is of general interest to the Group
because it raises all the major issues of the New
International Economic Order, and because it is
thought by many to have considerable precedential
significance.

4. The argument concerning the importance of the
freedom of military navigation turns principally on
two issues: the extent to which the security and
effectiveness of the strategic submarine force
requires submarine transit through straits (it being
given that any such transit should be submerged and
not subject to an "innocent passage" regime), and
the extent to which sound U.S. foreign policy and
expected U.S. naval capacity require and permit
peacetime "force projection" in distant waters that
are to be reached via unreliably free waters. The
argument concerning the importance of universal
treaty protection of military and commercial freedoms
turns principally on two more issues: the extent to
which such freedoms might be secured by other means
at reasonable cost, and the extent to which a treaty
per se is a significant force in restraining "juris-
dictional creep." (For the author's analysis of
these issues see the articles "Interests" and
"Choices," cited in note 1. The argument concerning
the adequacy of proposed treaty language turns
principally on interpretation of the freedoms being
"subject to the relevant provisions of the present
convention" (the Informal Composite Negotiating Text,
Article 58, and related dispute settlement provisions):
and it also involves consideration of the extent to
which environmental provisions and deep seabed provisions may further erode the traditional freedoms.
It is perhaps noteworthy that notwithstanding its

interest in protecting freedoms of navigation, the
United States has been willing to extend its coastal
jurisdiction seaward with respect to mineral resource
development of the continental shelf (the Truman
declaration), fisheries management (the Fisheries
Conservation and Management Act of 1976), and environmental protection (the Clean Water Act Amendments of
1977), and it has generally acquiesced in the expan-
sionist claims of other coastal states.

5.   In December 1969 the U.N. General Assembly
adopted U.N. Resolution 2574D (XXIV) declaring a
moratorium on exploitation of the deep seabed pending
the negotiation of an international agreement to
govern such development. The United States and 27
other countries voted against this resolution (with
28 more countries abstaining). The United States
does not, in any case, recognize U.N. resolutions as
legally binding; and the official U.S. legal position
(disputed by most developing countries) is that all
traditional high seas freedoms still obtain in the
deep seabed area. The pending U.S. deepsea mining
legislation is intended to be consistent with this
view. See, e.g., the discussion of the relationship
of domestic legislation to existing international law
in U.S. House of Representatives Report 95-588, Part
1, "Deepsea Hard Minerals Act," August 9, 1977, which
states, "The Bill [then HR 3350] does not intend or
purport to make any exclusive or sovereign claim to
the areas where the nodules are located nor does it
assert, as against the rest of the world, that the
U.S. or its citizens have an exclusive right to take
the nodules. The bill merely establishes a mechanism
to control the activities of U.S. citizens..."
(p. 24).

6.   Legislation advanced in the U.S. Congress has
provided for revenue-sharing royalty payments (to be
held in escrow pending the negotiation of a comprehensive law of the sea treaty). The legislation

explicitly contemplates and provides for reciprocal
agreements among mining states to assure security of
tenure. Related legislation has been discussed in
the British Parliament, the German Bundestag, and
the Japanese Diet. The framework which this emerging
pattern suggests amounts to a "minitreaty," but one
that attends less to developing country participation
than it might. (See the section below, "A Path Not
Taken.") It is fair to note, however, that although
a minitreaty arrangement could be structured to provide for equity among rich and poor states at least
as well as the proposed U.N. treaty, it is unlikely
that such an arrangement will be so structured. It
may therefore be said that equity interests would
likely be better served by a U.N. treaty, although it
is important to note that this judgment reflects an
assessment of political reality and not an inherent
weakness of a minitreaty approach. The judgment that
a minitreaty type of system would be more efficient
than the type being negotiated is based on an appreciation of market regulatory mechanisms on the one
hand and, on the other hand, consideration of the
proposed Authority's institutional arrangements,
stringent production controls, regulatory uncertainties, minesite allocation criteria, interest in
regulated prices, and moratoria provisions.

7. A relatively simple, decentralized licensing
regime is approximated by the combination of domestic
legislation such as that being considered by the U.S.
Congress and reciprocal agreements among mining
states. A centralized, but minimally conditioned,
licensing system is represented by the draft treaty
proposed by the United States in 1970 (see Draft
United Nations Convention on the International Seabed
Area, August 3, 1970). A variant of the parallel
system—essentially two different systems side by
side, with a minesite "banking" provision for
balance—is represented by the Revised Single Negotiating Text, U.N. Doc. A/CONF. 62/WP.8/Rev. 1/Pt. 1,

6 May 1976. A somewhat cleaner variant of the
parallel system that is also more balanced (in its
provision for simultaneous development of the two
sides of the parallel system) is represented by John
Norton Moore et al., "Revised Package of Amendments
for the ICNT Illustrating a Balanced Development
System," in Oceanography Miscellaneous, Part 2, pp.
223-238, House Subcommittee on Oceanography, Serial
No. 95-44, Washington: GPO, 1978). A variant of a
mixed system is represented by Jens Evensen, "Proposed Compromise Formulations." UNCLOS III Doc.
77-76231, 3rd rev., and Doc. 77-76619 (June 11 and
29, 1977). Possible variants of joint-venture
systems are many. Insofar as they are unitary without clear rights of access for state-sponsored
parties, they may be conceived as analogous to land-
based systems managed by state-centralized economies.
The principal focus of present negotiation in the
U.N. Conference has the superficial appearance of a
mixed system, but the operational effect of a
unitary system without clear and continuing rights
of access for state-sponsored parties (Informal
Composite Negotiating Text, U.N. Doc. A/CONF. 62/WP.
10, 15 July 1977).

8. This highlighting may appear to exaggerate a
pattern of U.S. concession. But it would be fair to
suggest that it does so only to the extent that the
United States's starting position is judged to be
unreasonable on the merits. In this respect it is
important to emphasize that prevailing international
law would have allowed and would still allow a considerably freer, more decentralized, and very much
less equitable system than that initially proposed
by the United States. In the author's view, the
pattern of U.S. concession cannot properly be discounted at all on grounds of ungenerosity or extreme-
ness in its initial position; for in relation to
prevailing law and practice, the U.S. position was
generous, conciliatory, and highly forthcoming

(although the tactics of its presentation left much
to be desired). If, by contrast, the standard of
reasonableness must be conformity with the "common
heritage" principle as understood by developing
countries, then a unitary system is an almost inescapable corollary; and the reasonable range of
negotiation would be confined to the terms of a
unitary system. One might suggest, therefore, that
it may have been the very conciliatory quality of the
United States's initial position—ceding the "common
heritage" principle—that, ironically, rendered the
basis for arguing the reasonableness of that position
less tenable.

9. See Ambassador George T. Aldrich, memorandum and
associated documents on "Consultations on the Possibilities of Simplification," distributed to "members
of the Hard Minerals Subcommittee of the U.S. Advisory
Committee on Law of the Sea and others interested in
simplification of the ICNT," November 29, 1978
(unclassified).

10. The degree of actual supremacy could theoretically
be limited by a system of judicial review; but the
ICNT system provides very little scope for such
limitation, and it provides for election of judges by
the Assembly in any case. Another approach to limitation involves the allocation of key functions (e.g.,
contract approval, rulemaking, and regulating as in
ICNT Articles 158(svi), 160(x) and 160(xiv)) to a
Council in which minority interests are better represented and protected. The negotiations clearly contemplate a Council, but whether it is to have effective responsibility for key functions and whether it
will be structured to protect minority interests
satisfactorily remain to be determined.

11. The proposed treaty would regulate _all_ minerals of
the deep seabed—not only minerals found in manganese
nodules. It would, however, establish a specific

regime only for manganese nodules (other sources being not yet commercially attractive). But it would require the proposed Authority to develop, and states to ratify, regulations for the exploitation of non-nodule minerals before any party could claim title or rights to such minerals. Given the virtually certain veto power of developing countries (and perhaps other interest groups) in the proposed Council, and given the need for ratification, the possibility of stale-mate in the development of non-nodule regulations would appear to be considerable. Such a possibility could indirectly or directly lead to a moratorium on non-nodule mineral resource development. (See ICNT Article 137(3) read in conjunction with suggested revisions to the ICNT at NG1/10 Article 150 bis(3).)

12. Some members of the Group of 77 would argue that the structure of the Council represented in the ICNT or the system of "parallel access" presented by the ICNT constitute important concessions. Indeed, some members of the Group of 77 would oppose such con-cessions. But upon serious examination, these are concessions of appearance, but not of likely operating reality. Indeed, in many cases the appearance of "77" concession itself disappears upon serious exam-ination. It should be noted further that since only the United States and a few developed countries now possess the necessary technology and financing for deepsea mining, some short-term concession to these realities is inescapable if there is to be seabed mining in the near term. The more important period of interest, and the more appropriate focus for an examination of concessions, involves the medium and longer term. Here the pattern of what amounts to major unilateral concession by the United States seems even more striking.

13. To date, the pattern of the evidence suggests that the patience of the Conference has been consider-ably greater than the patience of the United States.

It seems fair to suggest further that unless U.S.
policy were to give greater relative weight to
interests in the seabed regime and to traditional
American politico-economic values associated there-
with, the pattern should be expected to continue.

14. The draft Aldrich text referred to at note 9
above is largely consistent with this assessment.
With the exception of voting arrangements, it essentially incorporates redrafted versions of most of the
adverse provisions of the ICNT. The dilemma presented by detail-deferral arrangements in general is
that to assure protection of U.S. interests, the
representation and voting arrangements of the interim
detail development group must be structured to give
what developing countries would construe as excessive
weight to U.S. interests; hence it is unlikely that
developing countries would agree to such structuring.
And in the absence of such protection, the United
States would be in a highly vulnerable position. For
although it is theoretically correct that the United
States could sign but not ratify the proposed treaty,
the very act of signing would arguably change international law somewhat; the act of signing would oblige
the United States not to take actions inconsistent
with the proposed treaty prior to a vote on ratification (hence, the United States might bind itself
temporarily to an effective moratorium provision);
and the prospect of a signed treaty coming into force
through ratification by other states would very
significantly increase pressure on the United States
to ratify--and decrease the political viability of
unilateral action. The notion that the United States
could sign and then "wait and see" the detail developed
by an interim group at no significant cost is simply
misguided. To achieve a detail-deferral arrangement,
the U.S. would have to concede either important substantive elements in the treaty to be signed or
important elements of leverage in the interim arrangements, or both.

15. This is not intended to suggest that ratification would be an easy matter; but simply that nonratification must be rated an unlikely outcome. For symptoms of potential congressional difficulty in the ratification process, see, e.g., letter to Hon. Elliot Richardson from Senators Jackson and Hansen, as Chairman and Ranking Minority Member of the Committee on Energy and Natural Resources, dated August 2, 1978, and see letter to President Carter from Congressmen Murphy and Breaux, as Chairman of the Committee on Merchant Marine and Fisheries and Chairman of the Subcommittee on Oceanography, respectively, dated January 29, 1979. (Note: The U.S. House may have an important role in developing essential implementing provisions for a treaty, insofar as the treaty will raise issues of taxation and insofar as the treaty will require substantial appropriations for the subsidy of the start-up of the proposed Enterprise and Authority.)

16. For the author's assessment of these interests, see the articles "Interests" and "Choices" cited at note 1. For an analysis of the extent to which the proposed treaty language may support these interests, see Richard G. Darman and John T. Smith II, "Report of the Committee on Law of the Sea," Proceedings and Committee Reports of the American Branch of the International Law Association, 1977-78 (New York, 1978).

17. For pessimistic estimates of future manganese availability, see Robert L. L'Esperance, Statement submitted for the record of the Subcommittee on Oceanography, April 22, 1977, and Franz R. Dykstra, "Manganese--Its Strategic Implications," January 4, 1979. These estimates are disputed by the Department of Interior, whose estimates are much more optimistic.

18. The 1970 trusteeship proposal of the United States, which would have shared revenues from all

mineral resources beyond the 200 meter isobath, had
the potential for much more significant redistributive
effects. But that proposal has long since been
rejected by the Conference and would not now command
the necessary support for its revival (not least
because coastal states have, for obvious reasons,
become increasingly interested in directly controlling
and benefiting from the exploitation of the oil and
gas resources of the continental margin, defined as
broadly as possible).

19. See the third and fourth points of introduction
and the article "Precedential Implications," cited
at note 1.

DEVELOPING COUNTRIES' EXPECTATIONS FROM AND RESPONSES TO THE SEABED MINING REGIMES PROPOSED BY THE LAW OF THE SEA CONFERENCE

A. O. Adede

INTRODUCTION

The resources of the deep seabed are the manganese nodules, which have been described as small potato-shaped rocks generally found at depths greater than 2,000 fathoms on the surface of the deep seabed and deep ocean basins. The nodules come in different sizes, and their component metals also vary. The most important metals associated with the nodules are nickel, copper, cobalt, and manganese.[1]

According to the information available so far, the largest concentrations of nodules are found in the eastern Pacific Ocean, north of the Equator, with the most attractive sites in the region between Hawaii and Southern California. However, the manganese nodules are known to carpet the world's deep ocean floor. Thus certain areas such as the Indian Ocean where good grade nodules have been identified remain essentially unexplored. It is to be noted also that "while most of the nodule deposits are in areas beyond national jurisdiction, reports have indicated that some rich deposits may also be found within national jurisdiction, for example, near the French Polynesian Islands, Tonga Platform and Western Samoa. It is possible, therefore, that the international seabed mining system might have to compete with the national undertaking."[2]

The availability of manganese nodules has thus been determined, their metal components of principal economic interest ascertained, and the desire to explore and exploit them also established. It was on the basis of these facts that the United Nations passed a resolution which declared, inter alia, that the seabed and ocean floor and the resources thereof beyond the limits of national jurisdiction are part of the common heritage of mankind.[3]

Certain basic issues which have controlled the discussions on the formulation of treaty texts on the

subject of seabed mining may be noted at the outset.
First, emphasis is on the concept of the common
heritage of mankind and the notion that it is to be
exploited for the benefit of mankind as a whole. The
relevant resolution clearly established a particular
regime over the area.[4] Second, it must be borne in
mind that the nodules themselves lie, as noted
earlier, 2,000 fathoms down and that the requisite
technologies for their commercial exploitation are in
the hands of only a few multinational companies and
in the process of development,[5] thereby making deep
seabed mining all the more a capital-intensive operation. Third, the exploitation of the common heritage
of mankind should be undertaken through an international machinery properly constituted to take into
account the interest of mankind as a whole. The proposed machinery--the Seabed Authority--must be an
institution capable of exercising sound business
judgment consistent with the desire to maximize the
benefits from seabed mining for all mankind. In
order to do this, the machinery would be called upon
to assess the actual and potential negative impacts
of seabed mining on the economies of the land-based
producers of the same metals, the interests of consumers, and the needs of the developing countries.
     Throughout the seven years of consideration of
these issues in the United Nations Seabed Committee,[6]
and the present discussions in the Law of the Sea
Conference since the Caracas session of 1974, the
following questions have persisted, and the Conference,
now going to its eighth session, has struggled for
appropriate answers. The questions are: How should
rights of access to the seabed resources be allocated?
Who should carry out the seabed activities? How much
should be produced from the seabed resources? How
should the gains be distributed?[7] The developing
countries have made their views known on all these
and related questions and have articulated their
expectations from the seabed regimes being proposed
under the draft law of the sea convention.

As a result of evolving discussions during the
Seabed Committee and Law of the Sea Conference
sessions, certain assumptions on which proposals for
treaty texts were based have undergone critical
reevaluation and are in the process of being changed,
or even abandoned. A discussion of some of these
key assumptions is necessary for an assessment of
developing countries' expectations from and responses
to the regimes for the deep seabed so far suggested
by Conference documents.

DEVELOPING COUNTRIES' VIEWPOINTS ON DISPUTED POINTS
IN THE PROPOSED SEABED MINING REGIMES
     1. Reliance upon the nickel market as the yard-
stick. It has been generally acknowledged that seabed mining should be viewed primarily as a nickel
recovery operation. This assumption has made the
nickel market the yardstick for determining the
potential impact of seabed mining and has encouraged
the formulation of responses to proposed regimes on
the basis of suggestions made by the land-based producers of nickel. Thus subparagraph 1B(iii) of
Article 150 of the Informal Composite Negotiating
Text (ICNT) addresses itself to the protection of
the land-based nickel producers.[8] This protection is
now further refined in Article 150 bis prepared
during the seventh session of the Conference.[9] The
formula in Article 150 bis is essentially an insurance
policy for land-based producers aimed at avoiding a
rapid accumulation of seabed production which would
curtail the ability of such producers to recoup their
investments during the phase-in period of seabed
mining. It has been established that land-based
producers can supply enough nickel to meet the demand
for a number of years to come. Accordingly, from the
point of view of the consumers, nickel is not a
problem.
     What seem now to be the resources of concern are
cobalt (produced in Zaire, Zambia, Cuba, and Morocco)
and manganese (produced in Gabon, South Africa, the

USSR, India, and Brazil). The seabed text should
therefore focus attention on the protection of the
interests of the developing countries that produce
these metals. The desire of consumers for access to
the seabed resources as independent supplies of
cobalt and manganese[10] must be weighed against the
impact of such access to the economies of the land-
based producers of these metals. Developing countries
expect that the treaty should protect them and that,
where they are the land-based source of the metal,
they must be given due consideration by consumers.
A situation should not be allowed to emerge where
access to seabed resources is emphasized when the
land-based producers of the desired metal are developing countries, whereas access is controlled when the
land-based producers of the desired metal are overwhelmingly developed countries.

2. Profitability of seabed mining as the reason
for the desired access to the resources. There has
been a basic assumption that the seabed resources are
so vast and access to them so necessary that developing countries should stop obstructing the process of
concluding a treaty, and that the treaty should
embody the concept of "automatic access" or "assured
access" to those who have the capacity to exploit the
resources for the benefit of mankind. Now it seems
that the profitability of the seabed mining has been
exaggerated. Indeed, the developed countries have
now openly doubted its economic viability. One might
argue, therefore, that in supporting these doubts of
economic viability, the developed countries, or at
least some of them, must be understood as continuing
their interest in assured access basically for
strategic and political reasons. Thus the old assumption that access to seabed resources must be assured
for reasons of economic viability should be reevalua-
ted and the implications of the change in assumption
suggested here must be weighted to protect the
interests of all actors.

3. The value of the nodules and the bases for
financial arrangements for the Authority. It has
been a basic assumption of the developing countries
that the nodules in situ have a value which must be
paid for by those capable of mining them. The
developing countries have thus expected treaty texts
embodying financial arrangements reflecting the true
nature of the value of the nodules and emphasizing
profit-sharing between the Seabed Authority and the
contractors exploiting the resources. The developing
countries therefore question the principle of
"attributable net proceeds"[11] supported by the
developed countries, which, inter alia, gives the
nodules in situ a zero value. The other basis for
computing the share of the Seabed Authority in a
seabed mining operation would be the principle of
"net smelter return" or "net mine mouth return."
These competing principles for determining the share
of the Authority reflect varying degrees of implementation of the notion of fair return on investment.
The principle of fair return on investment finds
support in the proposals for financial arrangements
based on negotiated Discounted Cash Flow, which itself raises a number of unresolved questions.[12] It
would seem reasonable, then, to observe that the
assumptions on which the proposed financial arrangements for the Authority remain open to question.
Despite the painstaking efforts so far made in the
Conference, the use of the model for the preparation
of permanent treaty texts on the subject may not be
totally acceptable. There are just too many variables
about which we know too little in the context of
seabed mining to warrant the preparation of a mining
code.[13]

4. Treatment of land-based producers as a
cohesive group for special representation in the
Council of the Seabed Authority. In the efforts to
constitute the Council as an organ of the proposed
Seabed Authority, it has been assumed, for example,

that all the land-based producers of nickel, copper,
and cobalt would form a single special-interest group
to be represented in the Council. It is not clear,
however, that the behavior of land-based producers
will be the same in all situations. Since seabed
mining would most likely have significant reper-
cussions for cobalt producers such as Zaire, Zambia,
Cuba, and Morocco, it is difficult to see those
countries seeking the same protection under the treaty
as the developed countries that produce nickel, a
metal for which seabed mining has become less threatening. Approaches to protecting the interests of
land-based producers of the critical metals include
compensatory payments and commodity arrangements.
One problem with the approach of directly compensating
producers for losses arising from seabed mining is
that the revenues from the mining would thus be
depleted, leaving little to be distributed to the
international community. A number of methods have
been proposed for offsetting the loss of revenue by
the Seabed Authority through compensatory payments,
including the levying of an internal tax per ton of
cobalt consumed in the industrial countries.[14] The
use of commodity agreements is also not free from
difficulties. The success of such arrangements
depends largely on the ability of the developing
countries and other negotiators to get a reasonable
reading of future trends and developments by which
they can clearly assess their interests and possible
confluence of interests.[15] Failure of such readings
would lead to distorted assessments and a breakdown
in protection. Consideration of all these issues--
differences in the impacts of seabed mining on producers of different metals and the various approaches
for protecting the interests of the affected land-
based producers--argues against treatment of the
land-based producers as a cohesive group for the purposes of representation in the Council.

   5. The nature of the Seabed Authority: the
optimistic vs. the pessimistic view. On the question

of the international arrangements for the Seabed
Authority, the developing countries have maintained
that it is possible to create an intergovernmental
body unique enough to play the dual roles of supervising the implementation of the treaty obligations
in the seabed area and engaging in the actual exploitation of the seabed resources, in collaboration with
the other actors, for the benefit of mankind as a
whole. For this, the developing countries proposed
the establishment of the Enterprise as the operational
arm of the Seabed Authority.[16] The Enterprise was
conceived as a state-owned entity that would cooperate
with the other entities possessing seabed mining
technology for the purposes of exploiting the resources
of the seabed for mankind as a whole and not for the
benefit of the developing countries only. It was
expected to acquire managerial and operational
capability so that it could carry out the business of
direct exploitation of seabed resources without
necessarily competing with the other entities that
already possess seabed mining technology.[17] Unfortunately, the Enterprise has now come to be seen as a
sort of poor man's multinational competing with the
likes of Rio Tinto Zinc, Kennecott, and other
developed-country multinations.[18] This view of the
Enterprise should be corrected as part of the campaign aimed at establishing an optimistic outlook
about the Authority.

The developed countries have continued to maintain
a rather pessimistic attitude toward the proposed
Seabed Authority. They assume that the Authority will
be dominated by the developing countries through
irrational and nonvoting tactics, relying basically
on noneconomic criteria for reaching decisions.
According to the developed countries, therefore, the
Seabed Authority cannot be expected to be capable of
exercising sound business judgment in the carrying
out of seabed mining. Because of these competing
assumptions, the role of the Enterprise has not
become distorted. The developing countries expect
the status of the Enterpise to be reviewed, with

serious consideration given to their original argument
for its establishment. The regime proposed by the
Law of the Sea Convention should, in the view of the
developing countries, take an optimistic view of the
Seabed Authority, using the examples of existing
intergovernmental bodies which have so far success-
fully served mankind.

   6. Distribution of functions between the Assembly
and the Council of the Authority and the decision-
making formula for the organs. Still on the larger
question of institutional arrangements for the Seabed
Authority, it should be observed that the developing
countries have supported strengthening the Assembly
as the supreme organ of the Authority charged with
formulating policies for and giving guidance to all
other organs, including the Council, concerning
activities in the seabed area falling under the competence of the Authority.[19] The Assembly is regarded
by the developing countries as a plenary organ in
which all the parties to the Law of the Sea Convention
should be represented on the basis of equality, each
exercising one vote. Under this scheme, the Council
would simply be an executive organ supervising the
implementation of policies emanating from the Assembly.
The developed countries have, on their part, sought
to strengthen the Council and remove it from the grips
of the Assembly. Their earlier suggestions included
an attempt to assign to the Council the function of
laying down specific policies to be followed, thereby
sharing with the Assembly the power of policy-making.[20]
In order to elevate the Council, the developed countries have also favored the establishment of a number
of subsidiary organs, such as a Rules and Regulations
Commission and a Technical Commission added to the
Economic Planning Commission. The developing countries are quite concerned about this tendency to
expand the scope of the functions of the Council's
subsidiary organs to the extent of threatening the
competence of the Assembly on policy issues. Close

examintation seems now to favor resistance to the
proliferation of subsidiary organs in the Convention
and retention of a simple enabling clause calling
for their future creation as appropriate by a principal organ of the Authority. The notion is that the
Authority should not be cumbersome in its structure
at the outset. Thus, for reason of efficiency and
economy, only two subsidiary organs of the Council
need be established: (1) a Legal and Technical
Commission, and (2) an Economic Planning Commission.[21]
The distribution of powers and functions between the
Assembly and the Council and the composition of the
Council are important considerations in the decision-
making formula for these organs. While the developing
countries have now accepted the representation of
special interests in the Council,[22] it is doubtful
that they can accept any proposal that has the effect
of weakening the Assembly. Accordingly, a weighted
voting formula for the Assembly would be objectionable
to the developing countries, as would a voting formula
in the Council that constitutes a veto for the industrialized countries. By actively reexamining the
criteria for distributing powers and functions
between the two principal organs, and the qualifications for membership in the technical subsidiary
organs of the Council, and by supporting the sub-
ordination of the Council to the Assembly on questions
of policy, the developing countries seek to safeguard
their interests through the treaty.

    7. The question of basic norms for exploitation
vs. the elaboration of a mining code in the Law of
the Sea Convention. Consistent with their more
optimistic view of the Seabed Authority, the developing countries originally took the position that only
certain basic norms, capable of being reduced to
treaty provisions in the main body of the Convention
itself, should be formulated by the Conference.[23]
The promulgation of more detailed rules and regulations for seabed mining was to be left for the

Authority to undertake later on the basis of
experience and further technical developments. This
position was firmly supported by the developing
countries during the 1974 Caracas session of the
Conference.[24] It was, however, rejected by the
developed countries, which sought to remove from the
Authority all discretion with regard to the promulgation of rules and regulations for seabed mining.
They wanted to put the Authority in a straitjacket
from the start. Accordingly, the developed countries
supported the inclusion in the Convention of an annex
containing detailed rules and regulations for seabed
mining, amounting to a mining code or an insurance
policy that attempts to take into account every
conceivable situation.[25]

As discussions progressed in the later sessions
of the Conference, the developing countries abandoned
the idea of bare basic norms on the body of the
Convention and accepted the idea of an annex containing a set of limited yet detailed "basic conditions"
for the exploration and exploitation of the seabed.[26]
The developed countries also temporarily deemphasized
the idea of a mining code. But as the elaboration of
the "basic conditions" progressed, it became clear
that what was emerging was a de facto mining code.
The sixteen paragraphs of Annex II of the ICNT
entitled Basic Conditions of Exploration and Exploitation clearly embody more details than the developing
countries originally supported. To these are now to
be added a detailed specification of financial
arrangements for the Authority, currently being
worked out at the Conference. The elaboration of
financial arrangements has succumbed to the mining
code approach. The result so far is a detailed
document that relies on data and models yet to be
proven and tested. Accordingly, voices have been
heard again rejecting such detailed provisions on
matters about which very little is still known.[27]
The developing countries must now review the result.
The question is: Should we go back to the notion of

basic principles and prune the details out of these
annexes, leaving their formulation for further
scrutiny?

   8.   The sites for processing plants and protection
of the environment.   The seabed articles contain provisions calling for protection of the marine environment.[28]   Accordingly, it seems reasonable to expect
that environmental considerations will be taken into
account also in the processing of nodules mined from
the sea.   The assumption that the processing plants
will be located near the actual minesites will not,
therefore, hold true in every case.   It may well be
necessary to require a kind of an environmental
assessment statement before locating the processing
plant in a particular site.   Thought should be given
to the idea of treating mining activity as a trans-
action involving the Authority, the company or entity
that actually operates the mines, and a State--
possibly a developing one--which provides an appropriate site for processing the nodules.   This would
be another benefit accruing to the developing countries from seabed mining.

   9.   Settlement of disputes.   The developed countries have supported the establishment of a system
for the settlement of seabed disputes through a forum
with jurisdiction over all the would-be actors in the
area and having an extended competence ratione
materiae.   The developing countries, on the other
hand, while accepting the need for establishing a
disputes settlement system, assign very limited
competence ratione materiae to the envisaged forum.[29]
By supporting integration of the seabed mining disputes settlement procedures into the general system
for settling disputes under the Law of the Sea Convention as a whole, the developing countries opted
for abandonment of the idea of creating a Seabed
Tribunal as an organ of the Authority.[30]   This integration would result in a Seabed Disputes Chamber

within the proposed Law of the Sea Tribunal. The
developing countries would deny the Seabed Disputes
Chamber the competence to challenge discretionary or
legislative acts of the Seabed Authority in any
situations, including the process of promulgation of
rules and regulations or the awarding of contracts for
seabed mining. They do not want to.expose the
Authority to unnecessary harassment by applicants
through endless suits brought against the Authority.
Thus, judicial control of activities in the seabed
area still requires further examination aimed at
distinguishing cases of violation of substantive
provisions of the Law of the Sea Convention (the
question of "unconventionality") from cases arising
from administrative irregularities in the process of
dealing with applications for contracts, enforcement
of contracts, and the manner of implementation of
rules and regulations in specific individual
instances. The developing countries thus expect the
Convention to avoid including provisions for the
settlement of disputes that have the effect of eroding
the power of the Authority as the guardian of the
common heritage of mankind. The developed countries,
meanwhile, seek to ensure that, where the "guardian"
has acted in a manner inconsistent with the Convention and the rules and regulations made thereunder,
such actions should be the subject of judicial proceedings aimed at providing appropriate remedies for
the parties aggrieved by the action or omission on
the part of the Authority.

     10. Transfer of technology, not a purchase of a
cookbook. On the important question of transfer of
technology to the Authority, the developing countries
have maintained that such a transfer is an attribute
of the resource policy of the Authority which it may
invoke, among other policies, to determine the award
of a mining contract to one applicant over another.
The proposed Article 144[31] for the improvement of
the former paragraph 8 of Article 151 of the ICNT

addresses itself to this question. The Article lays
down requirements for transfer of seabed mining
technology from multinational companies to both the
Enterprise and the developing countries.

From the procedures envisaged under the proposed
treaty texts, it seems clear that the transfer should
not be designed to take place in such a way as to
amount to the purchase of a cookbook, whereby the
purchaser--the Enterprise or a developing country--is
left alone to follow the recipe without practical
assistance from the developer and user of the book.
The transfer of seabed mining technology should most
appropriately take place in the context of a joint
venture between the entities possessing it and those
seeking it. In this way effect can be given to the
proposed treaty text which provides, inter alia,
that "Measures directed towards the advancement of
the technology of the Enterprise and the domestic
technology of developing countries, particularly
through the opening of opportunities to personnel
from the Enterprise and from developing countries for
training in marine science and technology and their
full participation in activities in the area."[32]

The developed Western countries have rejected the
idea of creating a treaty obligation to transfer
technology in whatever manner. They point out that
the technology belongs to private companies that
cannot be legally bound, in advance, to take a
specific action with respect to such property rights.
We may thus be left with no other option but to
return to the joint-venture approach. The expectation of the developing countries is that the Convention should contain a provision aimed at achieving
the necessary access to the seabed mining technology
"under fair and reasonable terms and conditions."[33]

11. Benefits from seabed revenues vs. access to
seabed technology. Maximization of revenues from
seabed mining for distribution on the basis of criteria favoring developing countries is one of the

basic policies of the Seabed Authority. While the
relevant texts have always considered the distribution
of revenues and the transfer of technology on the same
level,[34] there seems to have been an assumption that
all developing countries would initially be interested
only in the revenue aspect. It should be borne in
mind, however, that some developing countries are
already in a position to weigh the benefits accruing
from the distribution of revenues against the benefits
arising from access to seabed mining technology on
fair and reasonable terms. Thus, as noted earlier,
access to seabed mining technology is clearly an
objective that developing countries expect to achieve
through the treaty.

12. The case for an interim committee to
elaborate basic conditions for seabed mining. At the
conclusion of the comprehensive Law of the Sea Convention, the immediate issue would be how long mankind is prepared to wait before commencing the
exploitation of the common heritage. For the sake
of the companies that were clamoring for immediate
access to seabed resources, earlier versions of the
seabed texts dealt with the question or provisional
application of the seabed part of the Convention.[35]
Further discussions disclosed resistance from those
who did not want to risk the provisional application
of temporary rules and regulations. Their objection
was supported by the common institutional tendency
to make permanent what has been once accepted as
temporary. Accordingly, the ICNT abandoned the idea
of provisional application of the Convention and
directed its attention to problems of periodic
reviews[36] and to a review conference for the entire
seabed regime after twenty years of implementation.[37]
The reluctance to allow provisional application
of the annex on the basic conditions for seabed
mining was an expression of lack of support for the
details, about which a number of negotiators felt
they knew very little. This issue is now to be faced
again and various options to be considered.

One option is to entrust to a committee of compe-
tent people the elaboration of rules and regulations
for seabed mining outside the politics of the Conference after the Convention is adopted. The mode
and timing of establishing such a committee is open
to various approaches. First, it might be constituted
soon after the Final Act of the Conference has been
signed by all the participants, thereby confirming
the text of the adopted Convention. Second, it might
be constituted from among the states that have signed
the Convention pending their expression of consent to
be bound. The number of such signatory states
warranting a convening of representatives into a
committee will have to be agreed. Third, one might
want to wait for the entry into force of the Convention before the technical committee is constituted.
This option of constituting a committee to elaborate
temporary rules and regulations for seabed mining
pending promulgation of permanent rules and regulations by the Seabed Authority clearly assumes that
there will be no such detailed provisions in the
adopted text of the Convention.
    If this option of a technical committee does not
commend itself, then we may have to reconsider the
provisional application of whatever basic conditions
the Conference may adopt in the Convention. The
developing countries expect that the Convention will
include a provision dealing with this question in a
manner consistent with their desire to protect the
prestige of the Authority as the guardian of the
common heritage of mankind, applying the norms
capable of giving effect to that concept as described
in the Convention.[38]

CONCLUSION
The foregoing discussion was aimed at indicating the
expectations of the developing countries and their
responses to the regimes being proposed by the Law
of the Sea Conference concerning seabed mining. The
following points merit emphasis here by way of
conclusion.

1.   There is great sumpathy with the idea that,
in the event seabed mining becomes economically
viable, the revenues accruing therefrom should be
distributed by the Authority using criteria that
favor the developing countries.

2.   From the information available, it seems that
seabed mining is going to be regarded as a source of
minerals required by certain consumers for strategic
and political reasons. Thus access to seabed
resources would constitute an independent source that
a particular consumer with seabed mining technology
can tap to avoid exclusive reliance upon the land-
based producers of the minerals in question. In this
connection, the nickel and copper supply is not a
problem, but that of cobalt and manganese is.

3.   In their expectation of benefiting from seabed mining when it becomes economically viable, the
developing countries continue to support the establishment of the Seabed Authority capable of exercising
functions suited for the exploitation of the common
heritage of mankind as a whole. The Authority is
viewed not as a monster, but as an institution upon
which mankind has bestowed confidence as the guardian
of the common heritage, taking decisions aimed at
balanced protection of the interests of both the consumers and producers of the minerals in question.

4.   Consistent with this view of the Seabed
Authority, it has been argued that the treaty ought
to allow the Authority the discretion to deal with
the details of seabed mining on the basis of experience. Thus the developing countries support the
inclusion in the Convention of only the basic norms
for seabed exploration and exploitation. Assuming
that the developing countries would hinder the
commencement of seabed mining, however, the developed
countries originally sought to include in the Convention a mining code in the form of an Annex to ensure
the commencement of seabed mining as soon as the
Convention is completed.

5.   The detailed provisions that have emerged for
the financing of the Authority follow the mining code

approach and now stand as a telling lesson. The
exercise indicates that it is premature to burden
the Convention with details based on data and models
whose bases are doubtful. In the context of the
financial arrangements, the developed countries seem
to have been convinced that a mining code is not the
answer. They now support the idea of pruning out the
details. They are accordingly taking the position
that the developing countries have consistently
defended with respect to the basic conditions.

6. Consistent with their view of the Seabed
Authority, the developing countries take the position
that the system for the settlement of disputes aimed
at achieving judicial control of the activities in
the seabed area should not unnecessarily erode the
power of the Authority. The system should, however,
be used for protecting the legitimate interests of
all the actors in the ocean space.

7. In order to allow early commencement of
activities in the seabed area, a preparatory committee
might be constituted to examine the question of rules
and regulations for seabed mining and other procedural
matters. The composition of such an interim body may
be a difficult point since it must be truly representative "politically" and also technical in nature.
It may be necessary in this regard to visualize an
interim body constituted along the line of the Council
of the Authority in terms of membership and decision-
making process.

NOTES

1. A recent study quoted the following figures for
metal content of the nodules: nickel 1.5%, copper
1.1%, cobalt 1.5%, and manganese 33%. See Adams,
"The Development of Manganese Nodules from the Ocean
Floor: A Long Term Econometric Analysis, 2 (Univ.
Pa. Paper No. 402, 1978).

2. Lee, "Machinery for Seabed Mining: Some General
Issues Before the Geneva Session of the Third United

Nations Conference on the Law of the Sea," in Proceedings of the Annual Conference on Law of the Sea (Gamble, ed., 1974), at 122.

3. See G. A. Res. 2749 (XXV), 25 UN GAOR, Supp. No. 25, Doc. A/8028, at 24-25 (1970).

4. See points 1-10 of resolution 2749 (XXV), cited in note 3, at 24. The same ideas are contained in Article 137 of the Informal Composite Negotiating Text (ICNT) dealing with Legal status of the Area and its resources reading thus (see Third United Nations Conference on the Law of the Sea, Official Records, vol. 8, 1978, at 23).

"1. No State shall claim or exercise sovereignty or sovereign rights over any part of the Area or its resources, nor shall any State or person, natural or juridical, appropriate any part thereof. No such claim or exercise of sovereignty or sovereign rights, nor such appropriation shall be recognized.

"2. All rights in the resources of the Area are vested in mankind as a whole, on whose behalf the Authority shall act. These resources are not subject to alienation. The minerals derived from the Area, however, may only be alienated in accordance with this Part of the present Convention and the rules and regulations adopted thereunder.

"3. No state or person, natural or juridical, shall claim acquire or exercise rights with respect to the minerals of the Area except in accordance with the provisions of this Part of the present Convention. Otherwise, no such claim, acquisition or exercise of such rights shall be recognized."

5. 1. Ocean Mining Associates (OMA) (A Virginia partnership). Consists of Deepsea Ventures, Inc., the service contractor; Essex Iron Company (a subsidiary of Union Miniere S.A. of Belgium); Tenneco

Corp. of the United States; and the Japanese Manganese
Nodule Development Co., Ltd.
  2. The Kennecott Copper Corp. Consortium. Consists of Kennecott Copper Corp. of the United States;
Noranda Mines of Canada; Consolidated Goldfields of
the United Kingdom; Rio Tinto Zinc of the United
Kingdom; British Petroleum; and Mitsubishi of Japan.
  3. Ocean Management Incorporated (OMINC). Consists of the International Nickel Company of Canada;
the AMR Group of West Germany; the Deep Ocean Mining
Company of Japan; and SEDCO Incorporated of the
United States.
  4. Ocean Minerals Company (OMCO). A partnership
of Ocean Minerals Inc., and Amoco Ocean Minerals Co.
Amoco Ocean Minerals is a subsidiary of Amoco Minerals
Co., Chicago, which in turn is a subsidiary of the
Standard Oil Company of Indiana. The Royal Dutch/
Shell Group member Billiton International Metals BV
of the Netherlands, BKW Ocean Minerals BV, a sub-
sidiary of the Royal Bos Kalis Westminister Group
N.V., and Lockheed Missiles and Space Co., a sub-
sidiary of Lockheed Corp., are the share holders of
Ocean Minerals Inc. A fourth member is being sought
to complete the consortium.
  5. The CLB Group. Consists of about 20 companies
from six countries, namely, France, Canada, Australia,
Japan, the United States, and West Germany.
  (Data from "A Resource Policy for Sea-Bed Mineral
Development: Issues and Implications" (not yet published, paper presented at a Seminar in Tokyo,
September 1978), Annex I.)

6. For a concise summary of the work of the U.S.
Seabed Committee, see, e.g., S. Ida, The Law of the
Sea in Our Time. II. United Nations Seabed Com-
mittee 1968-1973 (1977).

7. These questions are also raised in the Adams
paper cited in note 1, at 1.

8. See the U.N. document cited in note 3, at 25.

9. See Third United Nations Conference on the Law of the Sea, Official Records, vol. 20 (1978), at 23.

10. It is my understanding that during the recent debates in the U.S. Congress related to the pending bill on seabed mining, the concern was a lack of manganese and not the problem of nickel supply.

11. Compare the report of the Chairman of Negotiating Group 2 in the volume cited in note 9, at 68.

12. Some of the questions concern what constitutes a reasonable return on investment and the type of risks that must be taken into account, such as remoteness of the area of operation, weather problems, and technology.

13. See the comments in note 27 below.

14. See the Levy paper cited in note 5.

15. Ibid.

16. Submitted to Seabed Committee in Doc. A/AC.135/49, 1971.

17. The question of direct exploitation of seabed resources by the Authority is the issue on which disagreement persisted between developed and developing countries. The developed countries rejected this idea whereas the developing countries have equally steadfastly supported it. The result is a mixed system of exploitation now reflected in Article 151 of the ICNT, at 26, and refined in the new proposals in the volume cited in note 9, at 25.

18. See the listing of the companies in note 5.

19. Article 158 of the ICNT, at 28. No further discussion at the 7th session of the Conference.

20. This is now reflected in Article 160 of the ICNT, at 29, and later refined in the proposal of the Chairman of Negotiating Group 3 (see the volume cited in note 9, at 158).

21. See the suggestions of the Chairman of Negotiating Group 3 as contained in the volume cited in note 9, at 161-163.

22. For an analysis of the position of the developing countries on these matters see Adede, "Law of the Sea-- Developing Countries Contribution to the Development of the Institutional Arrangements for the International Sea Bed Authority," Brooklyn Journal of International Law, vol. 4 (1977).

23. The idea of basic norms was a compromise suggested by the delegation of Jamaica. Most of the developing countries, however, did not want the inclusion in the treaty of any rules and regulations.

24. See Adede, "The System for Exploitation of the 'Common Heritage of Mankind' at the Caracas Con- ference," American Journal of International Law, vol. 69 (31), at 46-47 (1977).

25. Their proposal, at Caracas, was a short seventeen- paragraph document circulated as U.N. Doc. A/CONF.- 62/C.1/L.7, in Id. at 172-173.

26. The following statement made by the Chairman of Committee I of the Conference is instructive. Delivered at the close of the resumed 7th session of the Conference in New York, it stresses the inter- relatedness of issues falling within the competence of Committee I on the creation of the seabed regime and the international machinery for exploiting the resources of the Area. The Chairman observed (see the volume cited in note 9, at 123-129):

"For the highly industrialized States,
having failed to obtain an absolute right
to naked veto, they seek refuge in detailed
provisions, as a source of that protection.
Written guarantees seem to be the motto.
The result is that we have not only plunged
our efforts into the undesirable elaboration
of a mining code for sea-bed mineral exploitation, but have ourselves been dragged into
adopting models and systems of calculations
on fictitious data that no one, expert or
magician, can make the basis of any rational
determination. We get more and more
engrossed with each session and have been
reduced to mere spectators in the inconclu-
sive tournament among experts.

"I would strongly recommend that we stop
this trend. I believe that this Convention
should encourage detail, but only such as
relate to matters that clarify the rights and
duties of States and identify the beneficiaries
of the common heritage. The package remains
one that gives assured access to those who are
capable of exploiting with the Authority for
the benefit of mankind as a whole on the one
hand, and on the other hand ensuring that
the Authority and its business arm (the
Enterprise) not only survive as a corporate
institution but also are adequately equipped
as viable ones with capability and ability
to fulfill the functions set in the
Declaration of Principles."

28. See Article 145 of the ICNT, at 24.

29. See Section 6 of Part XI of the ICNT, Articles
187-192, at 34, and Article 15 of Annex V, at 59.

30. For full discussion of this issue, see Adede,
"Prolegomena to the Disputes Settlement System Part

of the Law of the Sea Convention," New York University Journal of International Law, vol. 10 (253),
329-340 (1977, published in fall 1978).

31. See the volume cited in note 9, at 22.

32. Para. 2(b) of proposed Article 144, ibid.

33. Para. 2(a) of proposed Article 144, ibid.

34. See, e.g., Article 144 of the ICNT, at 24 and
the suggested improvements in para. 150(c) of the
document cited in note 9, at 22.

35. See Article 73 of the Informal Single Negotiating
Text (SNT) in the Third United Nations Conference on
the Law of the Sea, Official Records, vol. 4, at 148
(1975); Article 63 of the Revised Informal Single
Negotiating Text (RISNT) in Official Records, vol. 5,
at 136 (1976).

36. See Article 152 of the ICNT, at 26.

37. See Article 153 of the ICNT, at 26.

38. At this point one has to consider some of the
points made at note 27.

OCEAN MINING POLICY: A DYNAMIC APPROACH

Burton H. Klein

In order to better appreciate the policy issues
involved in coean mining it is necessary to under-
the payoff of such an operation from the standpoint
of society as a whole. Having first assessed the
deposits of nodules on the ocean floor, containing
manganese, iron, nickel, cobalt, copper, and various
other metals, one must then determine the potential
benefits of ocean mining to mankind. Such a determination demands consideration of the following
economic questions:

1. Is the world now facing a physical shortage of
such minerals?

2. Are the mineral prices rapidly rising?

3. If they are not currently rising, will mineral
prices rise in the future?

4. Is there an economic shortage of the minerals?

5. Is the main uncertainty over future supplies
political or economic?

6. What economic gains will result from mining the
ocean floor?

7. How will costs of metals taken from the ocean
floor compare to current or prospective costs?

Great uncertainty must be attached to the cost
estimates of new activities. Based on previous
research, it is doubtful that the costs in question
can be estimated within a factor of five or ten (that
is, the difference between the highest and lowest
costs depends in part upon luck, and in part upon
the cleverness of entrepeneurs during the design and
development stages).

Assuming that cost estimates exhibit a high degree of uncertainty, it would be appropriate to make a critical analysis of the participating firms by asking the following key questions:

1. Are the estimates made by comfortable monopolists with readily available financing?

2. Are the estimates made by eager monopolists who have become relatively risk-averse due to pressure resulting from operating within a consortium?

3. Or are the estimates made by highly enterprising and dynamic competitors who must take risks to assure their survival within the deep ocean mining industry?

It is important when viewing today's ocean mining firms to ascertain the extent to which their expectations are based on conservative design and the extent to which they are based on cleverness in minimizing costs.

Assuming that cost estimates for different nodule recovery and processing systems vary greatly (including present systems and potentially more innovative future systems), one must then determine the importance of minimizing costs associated with the utilization of the systems. One parameter in determining the importance of cost minimization is the elasticity of demand for various metals. For example, the high elasticity of demand for copper suggests that only a few cents difference in the price of this metal would make a great deal of difference in the quantity consumed.

If one assumes that deep ocean mining could yield significant cost reductions that would be important for the development of the economy, one finds that not only must there be a hidden hand at work, there appears to be an acute need for a hidden foot as well. Adam Smith's argument states that in making profits as large as possible, the producer serves not only his own interests, but the interests

of the public at large. The issue of technological
risk, as embodied in deep seabed mining, does not
change the essential character of Smith's argument.
However, Smith's definition of self-interest does not
include the risk of rivalry in the development of
minimum-cost alternatives. This is the hidden-foot
risk. The larger the risk your rival takes, the
larger the risk you must take if you value your survival. For example, it would pay to take larger
risks in the chemical industry than in the steel
industry.

Adam Smith cannot be blamed for having defined
self-interest so narrowly as to exclude the hidden-
foot concept from his analysis. In 1776, when he
wrote The Wealth of Nations, competition was ill-
organized, occurring between inventors, as opposed to
today's well-organized rivalry between firms. The
hidden foot began to play a progressively more important role in the making of significant advances
during and after the 1850s. In today's more dynamic
industries one finds that the hidden foot can be
deemed even more important than the hidden hand.
Though wide differences can be observed among
individuals with respect to their willingness to
engage in risk-taking, business firms seldom take
risks simply because they are headed by a heroic
entrepreneur. For example, when Henry Ford developed
the Model T, Ford's share of the automobile market
was only 10 percent, no larger than a dozen or so
other firms that had already gone out of business.
To compete against rugged Buicks selling for $1,000
and inexpensive runabouts selling for $400, there
was no alternative but to take the risk and make a
car that was as rugged as the Buick but would sell
for no more than $600. In other words, it was not
the hidden hand that was pulling Henry Ford; it was
the hidden foot that was pushing him. Indeed, in
many significant advances the push of the hidden foot
has been the primary motivation in inducing risk-
taking.

Having introduced rivalry into the scenario, one
must elucidate upon its role in reducing costs.
Assuming a subjective probability distribution with
respect to the cost of an operation (see figure 1),
one finds that while the firm may have a fair idea of
the distribution beforehand, it does not exactly know
where on the curve a particular hypothesis will lie.
If the firm is content with a hypothesis that falls
in the middle of the distribution (A on figure 1),
then by setting A as its limit cost it merely has to
search until it finds a satisfactory alternative.
This is the likely search strategy of a firm that is
not pushed to operate within a tight cost constraint.
Setting the cost limit at B, the firm will be provided with an incentive not only to search more
widely, but to ask more and more pointed questions.
The more pointed questions entrepreneurs ask, the
more they can reduce search costs. Finally, if the
limit is set at C, entrepreneurs are provided with an
incentive to extend the probability distribution.
The key point is this: the greater the change in
market shares that are likely to result from significant discoveries (that is, the more effective the
hidden foot), the greater the likelihood of extending
the probability distribution.

The entrepreneur cannot, however, foresee the
probability of extending the distribution. Indeed,
it is very important not to try to attach probability
estimates before making guesses about new hypotheses.
If probability estimates are attached to the hypotheses, many relevant hints will not be taken into
account. This new probability distribution is an
ex post facto distribution. For example, when the
Model T was first introduced, it cost $950. As a
result, its sales were disappointing. Consequently,
a $600 target price was set. Soon thereafter, an
employee of the Ford Motor Company informed Henry
Ford that he had visited a meat-packing plant and had
seen the disassembling of carcasses with overhead
production lines and questioned why it would not be

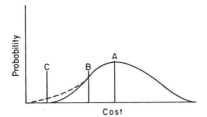

Figure 1. Operation X: Probability distributions
of hypotheses with respect to cost.

possible to reverse that process for assembling cars.
No one could have calculated the impact of this hint
on cost reduction. Ford implemented the idea where
its application would be most obvious. Only then was
the probability distribution extended.

In order to generate new probability distributions,
an organization resembling a criminal law firm is
required. But the organizations that operate with
mature technologies more closely resemble probate law
firms. A probate law firm trying a criminal law case
would encounter great difficulties. Analogously,
a highly structured organization (one set up for
specialized tasks) would encounter great difficulty
if it set out to develop new technologies. The
essential difference between an organization that
acts on the basis of given probability distributions
and one that generates new distributions is that the
latter must be far more interactive. The process is
often so interactive that the authorship of particular
inventions is in doubt. A highly interactive organization is one that encourages many lucky discoveries.
On the other hand, the more structured an organization,
the less likely it is that it can enlist good luck on
its side while at the same time minimizing the consequences of bad luck. Criminal law firms can become
married to and divorced from an entire succession of
hypotheses. Probate law firms cannot.

First-generation ocean mining firms will
undoubtedly subdivide in order to facilitate the
execution of various tasks. But if one reviews large
parent organizations of the past and present, one
finds that subdivisons are seldom given a great deal
of autonomy. For example, when a higher productive
dirm is acquired by merger, the parent organization
invariably imposes so many constraints that the
acquired organization can no longer remain innovative.
In the case of ocean mining, moreover, one must consider not only the constraints imposed by the parent
company, but also those imposed by other firms
belonging to the consortium, and those imposed by
various regulatory agencies.

As previously argued, it is impossible to imagine
a firm remaining dynamically efficient if it does not
face genuine competitive threats. The great degree
of dynamism demonstrated by Bell Telephone Laboratories after World War II, as compared to its performance in earlier periods, was the result of its
confrontation with real threats to its monopoly
position, including developments such as private
microwave stations, underseas telephone cables, and
communication satellites owned by other companies.
If a powerful hidden foot was required in the case of
Bell Laboratories, how can one expect dynamic
efficiency without a hidden foot in the case of ocean
mining?

If new ocean mining firms could expect the same
advantages in dynamic efficiency as new firms have
had in other industries, why have no independent new
firms entered the field? One possible explanation
has already been alluded to: the payoff for being
clever in ocean mining might not be nearly as great
as it is in other industries (a good example being
the semiconductor industry). A second reason is
simply the difficulties inherent in obtaining risk
capital. Raising a billion dollars for a venture of
this sort would be extremely difficult, and the cost
of entry is very high, particularly when compared
with other industries.

What, then, should the government do to encourage
competition? If minimizing the costs of these mining
operations is important, the government should share
the risk with newly founded firms. For example, the
government might provide the recovery system for the
firm in question. But the government should not pay
the entire cost of development, because that would
insure the new firms against failure. The firms
should be placed in a sufficiently desperate position
to be certain they will work to insure themselves
against failure. Second, it is questionable whether
one should give any firms, new or old, permanent
property rights for mining any specific part of the

ocean floor. The property rights should remain vested
in governments. On the basis of competition to
develop the lowest-cost equipment, the winning firm
should be given a contract of limited duration to
mine some part of the ocean floor and should also
reap the royalties if its equipment is used by other
firms.

Consider the railroads as an analogous system.
Suppose the government had owned the rights of way
and kept the rail system in repair. Also, suppose
that the government had then leased the lines on the
basis of competitive bids. It is not likely that the
second approach would have provided the railroads
with a far greater degree of dynamic efficiency. The
rate of diffusion of innovations in the railroad
industry has been slower than that of even the coal
mining or steel industry. Is it not conceivable that
if the government had leased the lines, progress would
have been more rapid? But what the government did
instead was to regulate the railroads in such a way
as to protect their right to be a cartel. This was
done under the guise of protecting the public interest.
But inasmuch as regulatory agencies prefer an environment with little uncertainty--because such an environment provides a way to minimize bureaucratic risk--
regulation cannot protect the common interest. Only
competition can do that.

One might argue that ocean mining regulation will
somehow produce entirely different results; but,
considering past trends, it is almost impossible to
believe that ocean mining regulation will prove to be
an exception. When firms cannot engage in risk-
taking, they either manage to set up an implicit or
explicit cartel or, failing that, they become regulated by the government. The only way to insure
against that kind of failure is to assure that the
industry in question is characterized by arduous
rivalry.

AN INDUSTRY PERSPECTIVE ON DEEP OCEAN NODULES

J. A. Agarwal

The resource classification of deep ocean nodules presents an enigma. When the ocean's vast wealth of nodules was first recognized, it was thought to be essentially a manganese ore resource "contaminated" with small quantities of copper and other nonferrous metals. Over the years the nodules were successively looked upon as a copper resource, a nickel resource, and finally, with today's high cobalt prices, a cobalt resource.

In evaluating mineral resources for economic development, one should attempt to answer the following questions:

1. Is the resource sufficiently large to warrant investigation?

2. Will it be economically exploitable?

3. Is it of strategic importance to the United States?

4. Will its exploitation help to minimize the balance-of-trade problem?

5. In what time frame can the resource be exploited?

6. What are the requirements for energy, water, manpower, transportation (from minesite to processing plant and from processing plant to consumer), and environmental control?

7. What effect will the exploitation of the resource have on the various related metal markets?

Ocean nodules represent one of the world's largest untapped deposits of nonferrous metals. Nickel, copper, cobalt, and manganese are present in

enormous quantities, enough to supply hundreds of
years of demand at current consumption levels.
Nevertheless, the nodules cannot be expected to dis-
place current metal resources until economically
competitive methods for the mining and processing of
the ore are developed. The technology required for
mining nodules from the ocean bottom, transporting
them to land, and then processing them are all highly
complex and extremely expensive.

In the case of processing, the problem is that
the nodule matrix is a mixture of iron and manganese
oxyhydroxides which cannot be easily beneficiated,
in contrast to sulfide minerals. The major metal
values--nickel, cobalt, and copper--are not present
as separate minerals but are distributed throughout
the oxide matrix. Nodules contain fine pores (100 A),
high porosities (60%), and a lot of water (40-50%).

Ocean nodules contain about 1-1.5% nickel,
1-1.5% copper, 0.1-0.5% cobalt, and 25-35% manganese
on a dry basis. The size of the mined nodules will
vary according to the recovery system utilized. The
delivered material may vary from a finely comminuted
sludge to whole nodule chunks. The nodule processing
plant must therefore be able to deal with a wide
variety of material as well as the presence of substantial amounts of sea water.

The first step in evaluating any mineral resource
is to determine the value of the contained metals
assuming complete recovery; such an evaluation for a
typical nodule ore is shown in table 1. Here copper
and cobalt are priced at $80/lb and $10/lb, respectively, which seems realistic in terms of long-term
trends.

Manganese stands out as the major component in
terms of value. But there would be heavy competition
from land-based manganese producers who are highly
integrated in the steel industry and utilize low-cost
ores grading 50% manganese. Furthermore, because
Pacific Ocean nodules are of higher grade than those
from the Atlantic Ocean, a U.S. recovery operation

Table 1
Metal values per dry short ton of ore

	Grade(%)	Amount(lb)	Price(%/lb)	Value($)
Manganese	30.0	600	0.25*	150
Nickel	1.3	26	2.10	55
Copper	1.1	22	0.80	17
Cobalt	0.2	4	10.00	40
Total				262

*Price in ferromanganese.

would most likely be on the Pacific Coast and therefore badly located with respect to the large U.S.
steelmakers. Consumption of manganese is keyed to
steel and has been growing at less than 4% annually.

The nickel contained in the nodules rates second
in overall value. Nickel has had a historic growth
rate of 6-7% annually. Despite the present over-
supply, the market for nickel is expected to continue
to expand because cost-effective substitutes for its
major applications are not readily available.

The degree of market penetration required of a
typical deepsea nodule mine is another important
consideration. The potential output of a 10,000 tpd
mine is shown in table 2.

The decision on the metals to be processed from
the nodules must be based on market demand. Because
markets may change, a process offering maximum flexibility in the selection of metal products is most
desirable. At present nickel is a prime recovery
target in deep ocean mining, but manganese will
increase in importance of terrestrial ores are
exhausted. A nodule processing venture that is
primarily oriented toward nickel production must,
however, be competitive with other on-line, land-
based nickel operations.

Another important factor in the analysis of
ocean nodules as a mineral resource is that the
metals contained therein, with the exception of

Table 2
Comparison of 10,000 tpd (3,000,000 tpy) nodule mine
output with estimated 1985 metal markets

	Mining Rate (short tons)	Percent of U.S. Use	Percent of World Production
Nickel	39,000	11	4
Copper	33,000	1	0.3
Cobalt	6,000	46	22
Manganese (all forms)	900,000	57	6

Source: Ocean Manganese Nodules, 2nd ed. Washington:
GPO, 66-675-0, February 1976, Manganese - 1977,
Bureau of Mines, MCP-7, October 1977.

copper, are for the most part not produced in the
United States at present. The risks involved in a
dependence on imports for strategically important
metals are clear.

Among the negative factors that must be taken
into account in the valuation process is the problem
of water disposal. For every ton of dry nodules, one
ton of salt water must be handled. Regardless of the
recovery process selected, the water content of
nodules will have an enormous economic impact.

Selecting a metal recovery plant site requires
answers to a host of questions whose economic impact
has been only dimly perceived until recently.
Regardless of the process selected, the plant will
require large amounts of energy, water, and manpower;
transportations systems into and out of the plant to
handle large quantities of materials and men; and
disposal facilities, such as land used to store tailings and slag in an environmentally safe manner.
Most cost estimates in the literature are unrealisti-
cally low because they do not adequately account for
the cost of transporting the nodules to an adequate
site.

There are two major types of processes available
for treating this complex mixture of metals: hydrometallurgical and pyrometallurgical. The complexity
and high water content of the nodules has led to the
consideration of several hydrometallurgical processes
in which the metals are solubilized from the nodules
by a strong leachant or by a combination reduction-
leach process. The leaching can be done by one of
the common mineral acids, such as sulfuric acid or
hydrochloric acid, or by a base such as caustic or
ammonia. East of reagent recycling is an important
economic consideration in these processes.
     The hydrometallurgical processes most commonly
proposed are:

1. solution reduction and ammoniacal leach,

2. high-temperature reduction and ammoniacal leach,

3. high-pressure sulfuric acid leach,

4. low-pressure hydrochloric acid reduction leach.

Overall, there is considerable flowsheet similarity
in the front-end material handling of nodule ore and
in the disposal of tailings for these alternatives.
There are also similarities in the methods used to
recover metals from the solution and in the final
production of copper, nickel, and cobalt. The major
differences are in the leaching system (including pre-
treatments to improve leachability) and in the method
of recycling reagents.
     In the pyrometallurgical processes, nickel,
copper, cobalt, molybdenum, and other metals are
recovered as a matte, whereas manganese is recovered
in a slag. Both the matte and the slag can be
further treated to recover the desired metals, ferromanganese and silicomanganese. Note, however, that
the nodule ore contains 30-40% moisture and is not
amenable to physical beneficiation. The entire ore

must, therefore, be dried and raised to smelting
temperature to recover less than 3% of its weight as
metal.

There is as yet no consensus on the best recovery
process. The Kennecott consortium has selected a
hydrometallurgical process, while the International
Nickel Consortium has selected a pyrometallurgical
process.

Table 3 compares the capital and cash operating
cost profiles for producing copper, nickel, cobalt,
and manganese from nodules by the hydrometallurgical
and the pyrometallurgical processes. To obtain an
unpurified manganese ore of similar quality to that
produced by the pyro process would increase the
capital investment of the hydro process by roughly
25%. Both processes require further purification of
the manganese ore to obtain a product suitable for
sale. The additional capital and operating costs
required for manganese recovery from the hydrometallurgical process tailings brings the two routes close
together, so the process choice must be dictated by
location, energy availability, and market strategy.

Manganese ore produced from nodules could be
further upgraded to form ferromanganese and ferro-
silicon products. The costs for the upgrading would
be approximately the same for the two processes.

Production and sale of manganese has a very large
positive impact on the profitability of the pyrometallurgical processes, but only a minimal impact on
the hydrometallurgical process. This results from
the fact that the slag produced in the pyrometallurgical process is already a dry impure manganese
ore, requiring only a low-cost purification step to
yield the synthetic manganese ore, whereas in the
case of the hydro process, the tailings slurry containing manganese must be upgraded by flotation, dry-
ing, and calcining before being purified.

Unlike the pyrometallurgical process, hydrometallurgical processes allow for the decoupling of the
production of copper, nickel, and cobalt from the

Table 3
Cost summaries for two alternative recovery processes

Products	Pyrometallurgical		Hydrometallurgical	
	Capital Cost Percentages	Oper. Cost Percentages	Capital Cost Percentages	Oper. Cost Percentages
Cu, Ni, Co	60	49	43	29
Mn ore (unpurified)	--	--	11	10
Mn ore (purified)	4	6	5	7
Cu, Ni, Co, Mn ore	64	55	59	46
FeMn and SiMn	36	45	41	54
Cu, Ni, Co, FeMn, SiMn	100	100	100	100

production of manganese, without adversely affecting
the economics, thereby allowing for a lower initial
investment.

Manganese nodules are primarily a nickel and
cobalt ore and offer better economic potential than
nickel laterites. The manganese ore has limited
market potential and is most desirable as an add-on
option independent of the production of nickel,
copper, and coablt. Therefore, a less capital-
intensive entry into ocean nodule processing strategy
can be followed.

Having examined nodules as a potentially valuable
resource and looked at some of the engineering and
marketing considerations governing their economic
exploitation, we should also evaluate the various
parameters a private company must consider in making
a systematic investment decision regarding the exploitation of a new resource. These parameters include:

1. present reserves and the value of the ore;

2. impact of a large-scale operation on the international and domestic market for the product;

3. cash required and time schedule for expenditure,
financing arrangements, return expected, and risks
involved;

4. legal issues (security of ownership, taxes);

5. time horizon for the project and the potential
impact of competition within the ocean mining
industry;

6. technology required and risks in its application
to the new resource.

Many of these parameters are interrelated, and
their consequences must therefore be analyzed in an
integrated fashion. It is safe to say that the

resource is very large and has substantial value.
Additionally, because of the high cost of ocean
mining, a multimillion ton per year venture would be
required to exploit the nodule resource economically.
Thus, any nodule mining ventures will have a potentially significant impact on the market for the metals
mined.

A 3 million dry short ton per year nodule project
would cost approximately 1 billion dollars, plus or
minus a few hundred million depending on the site and
the form of manganese produced. At these levels of
capital expenditure and at the current price level
for the various metals contained in the nodules, the
chances of receiving even a 15% DCFROR (the minimum
acceptable return) are slim. Presently the financial
health of the large companies most likely to participate in such a venture is not very good. Furthermore,
project financing may be difficult to obtain at a
time when the product (i.e., manganese nodules) seems
unattractive. And there also remain legal issues
relating to security of tenure and taxation policies.

The most important engineering issue relates to
the transportation of ocean nodules containing 40-50%
water to the processing plant. At a minimum, it is
necessary to transfer at least one ton of water for
every ton of dry nodules, although the quantity of
water that must be handled is considerably higher (by
a factor of 5 to 10) if slurry mining is considered.
In addition, the ocean floor is not sandy and granular
in substance, but rather contains a large fraction of
slimes that add to the bulk that must be handled.
Processing at sea has been considered, but no one has
solved the problem of how to supply the enormous
quantities of energy and chemicals needed on the ship
for processing.

Assuming a land-based processing plant, one
encounters another transportation problem resulting
from the distance from the plant to the product
market place. The West Coast of the United States or
Hawaii may not be good locations for a processing

plant if manganese is a major economic factor,
because most of the U.S. users of manganese are in
the Chicago and Pittsburgh areas.

In selecting a suitable minesite, many interrelated considerations must be examined. Along with
the transportation problems previously mentioned,
the engineer must consider the following:

1. site topography and acreage,

2. seismic activity and its impact,

3. availability of water,

4. availability of energy,

5. disposal of tailings and slag,

6. manpower availability and housing,

7, environmental laws and their impact on industry's
ability to operate (i.e., permits, licenses, etc.),

8. weather.

Ocean nodule processing plants are energy-
intensive operations. Industry has found that it
will require long lead times and various trade-offs
to procure the 60-100 megawatts and the other forms
of energy required for processing operations. For
example, in evaluating the advantages of a Texas
process plant location, industry must weight the
advantages of good supplies of energy and low seismic
activity against the necessity of transporting the
nodules through the Panama Canal.

If the right approach is taken, at-sea processing
may be a feasible alternative. However, it would
require a process design that operates under the constraints of energy supply on the high seas. It would
demand a radical change in the thinking of process

development engineers and would necessitate that they
work closely with ocean engineers.

In summary, ocean nodules are a large and rich
resource,containing many metals of great importance
to the United States and the world at large, but
prevailing unfavorable market conditions appear to
present a major deterrent to the economic feasibility
of deep ocean mining.

John E. Flipse

In analyzing the present deepsea mining operations
one may draw the following conclusions:

1. The United States must retain its right of
access to seabed nodules. The importance of right of
access may be displayed in two simple analogies. In
the mid-1950s the Newport News Shipbuilding & Dry Dock
Company made its paint in a company paint factory,
and its marine steam turbines by hand in the company
shop. The company was convinced by outside interests
that it was unwise not to buy "superior quality, formulated" industry-made paint. On competitive bids
the industry-made paint came in a dollar or two lower
per gallon than the company's paint. Upon evaluating
the situation, Newport News Shipbuilding & Dry Dock
Company closed its company paint factory. It was
only a short while thereafter that commercial paint
prices rose steeply.

Newport News Shipbuilding & Dry Dock Company also
made large marine steam turbines by hand. Of the two
American manufacturers only one was competitive in
price and delivery. In the mid-1960s when the company stopped producing turbines, the price of turbines advanced sharply, while the delivery time
changed from a year and a half to four years.

These analogies have been drawn to emphasize that
one must approach the concept "if you give up control,
the other party will treat you right," with skepticism.
The U.N. argument over "access to the area" is one of
control. The current debate in the U.N. is not a
debate over seabed resources but is, to the contrary,
an ideological battle over "The New Economic Order."

2. The U.N. Law of the Sea Conference is a
disaster. Henry Kissinger used the Law of the Sea
Conference as a grandstand, with devastating results.
Never once in the conference did the United States

receive a quid pro quo for a concession that it made.
Under the present circumstances, delay in ocean mining
is a clear victory for the LDCs.

Upon analyzing the current Informal Composite
Negotiating Text (ICNT) and recent U.N. negotiations,
one finds that:

--The 20-year review clause makes acceptance
impossible, since it requires renegotiation of a
"satisfactory treaty" or loss of all rights, which
is hardly the basis for an equitable end product.

--Every attempt made to provide a balanced renegotiation base has been defeated.

--The antimonopoly concept is an impossible restraint.
The conference is destined to be a disaster because
the values being negotiated are often of an emotional
nature.

Developing countries are worrying about face,
about righting prior wrongs, and about other values
that have nothing to do with resource development.
The language and the structure of the treaty
were intentionally designed to confound and confuse,
to make meanings obscure, and to make the treaty
subject to broad interpretation.

Such a design is clearly not in the interests of
the United States or industry. No treaty can solve
basic problems between parties if those parties do
not share a mutual trust, just as no contract can
make a venture succeed if the objectives of the
parties are not the same, much less diametrically
opposed.

3. U.S. domestic legislation must pass. If that
legislation passes, the Law of the Sea Conference
delay will result in a defeat for the LDCs as opposed
to a defeat for the United States. The ocean mining
bill HR 3350 was the product of several years of

compromise, had little direct value for the emerging
ocean mining industry, and as a result was viewed by
many members of that industry as basically useless.
On the other hand, the Senate version of the bill,
which was supported by Ambassador Elliot Richardson,
was a step in the right direction. It was very
unfortunate that Senator Abourezk chose to single-
handedly kill the Senate bill.

The important contribution of U.S. legislation
will be its stimulation of other foreign ocean mining
legislation, which is presently waiting in the wings.
Although excised from our legislation, bilateral and
multilateral treaties will rapidly develop on passage
of a U.S. ocean mining bill. After such passages one
will see a "departing-train syndrome" having the
slogan: "Ocean mining is starting. Either get on
the train or miss it." At that time the U.N. Conference on the Law of the Sea will end, and out of
necessity, real negotiations will begin.

The Law of the Sea Conference is dangerous in
that it draws time, money, and talent away from ocean
mining and other related areas. If one wonders why
the ocean mining advocates are not out advocating, it
is because one cannot advocate a dead horse. The
industry is a dead horse until it is backed by legislation. Although a potentially essential activity
of the future, ocean mining today is not a high
legislative priority.

4. <u>Deep ocean mining is commercially feasible</u>
<u>within five years if the following occur:</u>

a. The legal and political uncertainty is
eliminated. It is apparent that the U.S. legislation
will provide the solution in the elimination of this
uncertainty.

b. The private contractor has a minable
deposit. There are many minable deposits below the
high seas which can be located fairly easily and
evaluated for economic development through a

well-planned exploration program. In analyzing the
figures of Archer, Pasho, the French group, and
others, one must keep in mind that they all have very
strong motivations for concluding that there are only
a few minable deposits. A minable deposit is not
only characterized by the abundance and grade of
manganese nodules, but also by the terms and conditions of access.

   c. The private contractor has leveraged
capital. It is impossible for commercial ocean
mining to be financed with internally generated capital. The capital requirements are just too great.
The problem is that the return on investment is based
on one's equity, and that common stock price/earnings
ratio moves directly with the return on equity. The
future of a company chief executive moves directly
with the price/earnings ratio, and as a result there
will never be ocean mining without leverage capital.
One must have earnings and the euphoria it produces
in the parent company to support a novel, high-risk
program.

   d. The private contractor must have a
healthy market. The copper market price slump is
caused in part by Chile. The reorganization of the
Chilean government a few years ago required foreign
exchange, and Chile's only significant source of hard
money is its copper exports. Chile is therefore
flooding the copper market. It is the core of the
copper price problem, which is essentially a short-run
problem. Some other countries have also been selling
copper because they need the cash.

   The price of nickel is also low. The
nickel laterite mines are in a belt near the equator,
an area of intense political unrest. It would appear
that these mines are being "stripped." In other
words, they are being mined at a rate above the long-
term planned rate, because of the political uncertainties of the region. Such a strategy would provide the best chance of recovering one's investment.
Another factor in the low price of nickel is the

grade problem, present in all Canadian mines (with
the exception of Sudbury). It was disappointing that
the INCO consortium decided not to proceed with
commercial mining after its successful at-sea system
test. The politics of initiation of ocean mining
while reducing terrestrial mining is very sensitive
in Canada.

The price of cobalt has been artificially
set. It will be equal to nickel if and when there is
an independent major source of cobalt. The current
high price is a result of recent disturbances in
Zaire and cannot be considered in long-term
projections.

Looking toward other metals one finds
that the French 200-mile Exclusive Economic Zone has
an almost unlimited supply of 1% titanium assay manganese nodules. With this insight it is easy to
understand why the French changed their point of
view on the antimonopoly clause of the UNCLOS negotiations. The molybdenum and the zinc, although not
major constituents, are also interesting. Some con-
tend that there will be technological advances that
will permit industry to extract more and more metals,
of which there are approximately 27 in manganese
nodules, if processing research is approached from an
innovative viewpoint.

e. The private contractor has corporate
leadership. The leadership problem is intensified
by the normal progress of business and the changes
it produces.

For example, Kennecott Copper Corporation
recently survived a merger and now has a new Chief
Executive Officer. Such questions arise as: How
can a new Chief Executive, whose ultimate responsibility is investment decisions, justify spending
money on a "way-out" development like ocean mining?
Will U.S. Steel start commercial ocean mining before
the manganese mine in Gabon closes?

The Gabon mines will not be depleted for
twenty years, but could close long before then for

political reasons (there is currently a Cuban presence along the three borders of Gabon). It seems
apparent that INCO will go into ocean mining when it
is necessary, but there is some question as to what
business and leadership factors will prevail. There
is also concern over the oil industry's involvement
in deep ocean mining. Its attempts to diversify are
real, and several of them have very significant
mineral departments. There is some question as to
the oil industry's staying power, especially if it
encounters earnings problems. The leadership in
Japan is tremendously effective, especially in the
context of "Japan, Incorporated." but faces severe
supply problems, including international fisheries
which are of a higher priority than seabed minerals.
        f. The ocean mining industry must have the
national will in order to succeed. There is nobody
in the ocean mining business who doubts that existing
and pending government regulations can cripple operations when they are undertaken. In fact, government
regulations can stop anything! Similarly, misuse
of the courts can substantially delay anything, and
in terms of making this level of investment, would
also stagnate industry.
        One should adopt the "Orphan Annie
Principle." As Daddy Warbucks told Orphan Annie:
"Don't do good for anybody who doesn't want to be
done good for." For American industry to develop
deep ocean mining on a commercial scale, the investment decision-makers must see firm evidence of a
national will to proceed enunciated by both the
executive and legislative branches of our government.

        The alternative course of this kind of develop-
ment would come in response to a crisis. A general
war is not such a crisis. Stockpiles will take care
of a short war, while setting priorities and industrial mobilization can handle most materials problems
in a longer conflict. Ocean mining is not a realistic
approach to meeting long-run metal demands during

wartime. A real materials crisis could be generated
by the invasion of Gabon, the fall of South Africa to
foreign interests, or the rigid conservation of
national resources, as enunciated by Brazil and
Venezuela.

    5. The overall conclusion is that most of the
requirements listed in paragraph 4 must occur simultaneously if the United States is to have a successful commercial deep ocean mining industry. These
sets of circumstances were once considered to be
"fortuitous coincidence," but hopefully this is not a
prophetic observation.

    Let us now consider the economic feasibility of
deep ocean mining. Ocean mining is not now a "hot
opportunity" for investors. Its prospects have
changed significantly since 1968 when Gardiner
Symonds and Tenneco entered the field. At that time
a five-year research and development program at a
cost of about ten million dollars was proposed. The
program presently is in its eleventh year, and the
total investment is approaching ten times the original forecast. Of course, young entrepreneurs set
difficult targets and if somebody had actually wanted
to go ocean mining, a reasonable technological job
could have been accomplished at the end of the fifth
year.
    The U.S. conglomerates and the Japanese trading
companies have gotten out of ocean mining. The
technology developers, the metal companies, and the
"big oil" companies are in. The people who need the
feedstock are staying, and most risk-taking investors
start from "dollar one" and "day one" in evaluating
a major undertaking on a discounted after-tax cash-
flow basis. When using discounted cash flow as a
return on investment criterion, an interference
arises as a result of the amount of time consumed by
ocean mining R&D programs. Hence, the feedstock
approach is quite logical. INCO's nickel, Kennecott's

copper, and U.S. Steel's manganese concerns in the
long-term are clear. The Belgians have had a serious
raw material problem since they lost the Congo. The
big oil companies have the money and want to diversify, but will they stick it out? The big oil role
is questionable unless the companies are committed to
a minerals or metals business and need feedstock.

French Polynesia is one special situation that
is very interesting. CNEXO recently denied finding
manganese nodules in the 200-mile Exclusive Economic
Zone of the French Polynesian archipelago. Evidence
shows that this area has an abundance of potential
minesites. The nodules are of a high cobalt and high
titanium assay. Some hypothetical questions that
must be considered are: When is French Polynesia
scheduled for political independence? When does it
gain its economic independence? What is "Le Nickel"
position in deep ocean mining and in laterites?
What is the future of "Le Bomb" site?

Perhaps the last feasible approach would be
government ocean mining. A horrible thought, but if
the national interest was sufficient it might happen.
The vehicle could be subsidies or joint ventures, or
if our national security was threatened, the government could undertake deep ocean mining directly,
thereby eliminating many of the requirements presented
above. Government ocean mining is a possibility,
particularly in other countries.

In reviewing deep ocean mining technology, one
finds that the technology exists, with qualifications.
There is no question that Deepsea Ventures' and
SEDCO's at-sea operations have been successful. It
is likely that Lockheed's at-sea program will also be
successful, because of the loose definition associated
with research success. All consortia have most of
the basic technical information they need to develop
commercial ocean mining equipment, but the consortia
still face expensive engineering development and
tests. The shortcomings of these tests will identify
several areas of scientific investigation that must

be performed to optimize and reduce risk and/or cost
of the systems.

There are also a few questions in the area of
the processing technology. Manganese nodules are a
unique ore. As an entrepreneurial group with only
money (U.S. direction) coming from corporate supporters, the consortia were unrestricted and able to
explore unique processes. The most devastating
effect of heavy sponsor-corporate guidance is the
forced introduction of safe, proven, corporate
technology. Its adoption is fatal to the innovative
approach. The INCO-led consortium suffers from this
stagnation, especially when one considers the high
processing competence of at least three of its members. As an aside, let us hope that Lockheed builds
an innovative pilot plant in Hawaii. The technology
exists and will work, but there are still opportunities in processing research and development.

In conclusion, there are many valid reasons to
mine the deep ocean manganese nodules. The obstacles
are unfortunately many and varied. A real concern
is: Can a commercial ocean mining undertaking be put
together in a free country under a competitive
market? If it is not done in this manner, it will
be done some other way.

PART IV

SUMMARY AND CONCLUSIONS

POINTS OF CONSENSUS AND POINTS OF CONTROVERSY

Judith T. Kildow

The papers presented in this volume have offered a
variety of political and economic perspectives on the
seafloor ferromanganese deposits. Most of the papers
provoked lengthy and sometimes heated discussions
among the thirty or so participants in each of the
four seminars. I shall attempt in this chapter to
highlight the key points made in the papers and to
sketch some other points raised in the discussions.

The ultimate purpose of the seminars was to
determine a net strategic and economic value of the
manganese nodules to the United States. Several
components of the problem were assessed:

1. abundance and grade of the resource;

2. market structures for the relevant metals;

3. strategic problems attached to alternative sources
of the metals;

4. other qualitative sources of value (such as the
need to maintain a position as a technological
leader);

5. other major policy issues closely related to
deepsea mining (such as the general problem of the
proper economic relations between the industrial
nations and the Third World).

There seemed to be general agreement on estimates
of abundance and grade for the mid-Pacific, but there
was wide disagreement on other areas and also on the
rate at which the resource might be exploited given
better technology, improved exploration data, and a
more certain political and legal system. It is interesting that the number of mine sites, which has been
the unit of measure most commonly debated in the past,

was bypassed in the discussions for estimates by tons
for the nodules. In general the estimates presented
by industry representatives far exceeded those presented by academic ocean researchers.

The possible effects of deepsea mining on world
metal markets proved quite controversial. Clearly
there is a potential problem, since projected recovery rates for some of the metals indicate that they
could satisfy a sizable share of present world demand.
Some participants argued, however, that one must consider the possibility that the legal regime ultimately
endorsed might establish production and country controls. Others noted major sources of uncertainty in
projections of future demand. One paper, for example,
suggested broad new uses for manganese. Other participants cited substitution effects, other new uses,
and the effects of economic recession as factors that
must be taken into consideration in projecting demand.
The manganese situation was of particular interest
since U.S. Steel estimates of reserves differed
greatly from those of the U.S. Geological Survey.

Discussions of the strategic importance of the
deepsea resource for the United States also produced
a wide range of opinions. Concentrating on manganese
and cobalt, several participants built scenarios in
which the current sources of these metals became
unreliable. Others disagreed strongly, arguing that
stockpiling, cultivation of alternative suppliers,
and political and economic leverage still provide
more than adequate control over the security of
supplies.

Given that deepsea mining is economically justifi-
able and strategically advantageous, there remains
the problem of the proper auspices under which the
technology necessary for the mining should be
developed and deployed. The political problems,
involving as they do national perceptions and pride,
proved even more controversial than the economic and
technological problems. On the one hand, industry
representatives argued for the value of the industry

itself to the United States as a source of technological innovation and jobs and as a means of
stimulating the economy, improving the country's
international economic position, and guaranteeing
access to the resources. Third World countries, on
the other hand, see the development of the resource
as a means of initiating a fundamental change in the
world economic order. They argue that ownership of
the resource should be as serious an economic consideration as ownership of the means of exploiting
that resource. Since in this case ownership of the
resource is generally agreed not to be vested in
particular states, they argue that the economic
benefits derived from both the development of technology to be used in exploiting the resource and the
actual exploitation should accrue to all nations,
although proportionally more to the disadvantaged
nations to help offset the effects of past inequi-
table conditions.
    There were a number of gradations of opinion on
the section of the current negotiating text of the
U.N. Law of the Sea treaty that discusses the creation
of a seabed regime to manage this resource. At one
extreme were those who thought that the current text
as a whole represents a total loss as far as the
industrial nations are concerned and that these
nations should pursue other international arrangements for development of the resource. Slightly
differing were those who thought that the treaty as
a whole might be acceptable but that the provisions
regarding deepsea mining are not. This group
emphasized the inhibiting effect the proposed regime
would have on international resource allocations and
the bad precedent it would set for the control of
other transnational resources. They felt that the
resource is sufficiently valuable that the industrial
nations should pursue it on their own, but held
that this pursuit would have little effect on the
overall treaty negotiations. The next gradation of
opinion held that the industrial nations should

proceed outside U.N. auspices, but principally as a
negotiating tool toward an eventual compromise with
the Group of 77.

These views all reflected a position that nations
with mining technologies and vested interests should
have weighted control over any seabed regime. The
supporters of the regime proposed by the current
text also manifested a range of viewpoints. First
there were those who felt that the resource is sufficiently unimportant that the issue of its control
should not be allowed to inhibit progress toward a
treaty agreement that will be an important factor
favoring world order. Then there were those who
argued that the industrial world must capitulate to
the Group of 77 because the cost of not doing so
would be too high (the source of this anxiety was
never clearly brought out). Finally, there were those
who argued that change is more important than
economic efficiency, that it is in the interest of
the industrial world to foster the necessary changes
in the economic groundrules for the international
economic order.

A related question involved the necessity and
possible form of U.S. legislation to govern deepsea
mining. The range of opinions here mirrored those
expressed on the subject of the U.N. treaty.
Opposing legislation were both those who felt that
the resource is of no immediate consequence and those
who felt that the U.N. negotiations are too important
to be tampered with. Supporting it were those who
felt that U.S. legislation will catalyze positive
results from the United Nations and those who felt
that, even if a treaty does come about in time, it
is important to provide interim measures now to
bolster the interest of the industry and encourage
its technological investments.

The content of the legislation was also debated.
Industry representatives strongly advocated the need
for exclusive rights over mine sites, whereas
international lawyers and diplomats held that this

would be impossible. Other topics debated included
the relevance of the Jones Act, which requires U.S.
registration for ships transporting metals from mine
sites to U.S. shores. The focal issue of the debate
was the extent to which the United States can impose
legislative restrictions before they become counter-
productive, in that they drive the industry out of
the country onto more receptive shores. The thin
line between the necessity for control and the need
for incentives was left indistinct.

Two things became clear by the end of the final
seminar. First, the issue is uncommonly intricate
and tangled, mixing problems of resource supplies
and security, economic and political policy, in ways
that will not be easily or quickly sorted out.
Second, though, and perhaps more important, no matter
how uncertain or hostile the international legal
system may currently appear, the industry will,
sooner or later, launch production, although many
of the companies now heading in that direction may
not prove strong enough to outlast the turmoil
and enter the competition.